THE PRIVATE PILOT'S LICENSE PROGRAM

BEN P. MILLSPAUGH

TAB BOOKS Inc.
Blue Ridge Summit, PA 17214

FIRST EDITION
FIRST PRINTING

Library of Congress Cataloging in Publication Data

Millspaugh, Ben P.
The private pilot's license program.

Includes indexes.
1. Private flying—United States. 2. Air pilots—Licenses—United States. I. Title.
TL721.4.M53 1985 629.132'5217 85-22188
ISBN 0-8306-2371-X (soft)

Contents

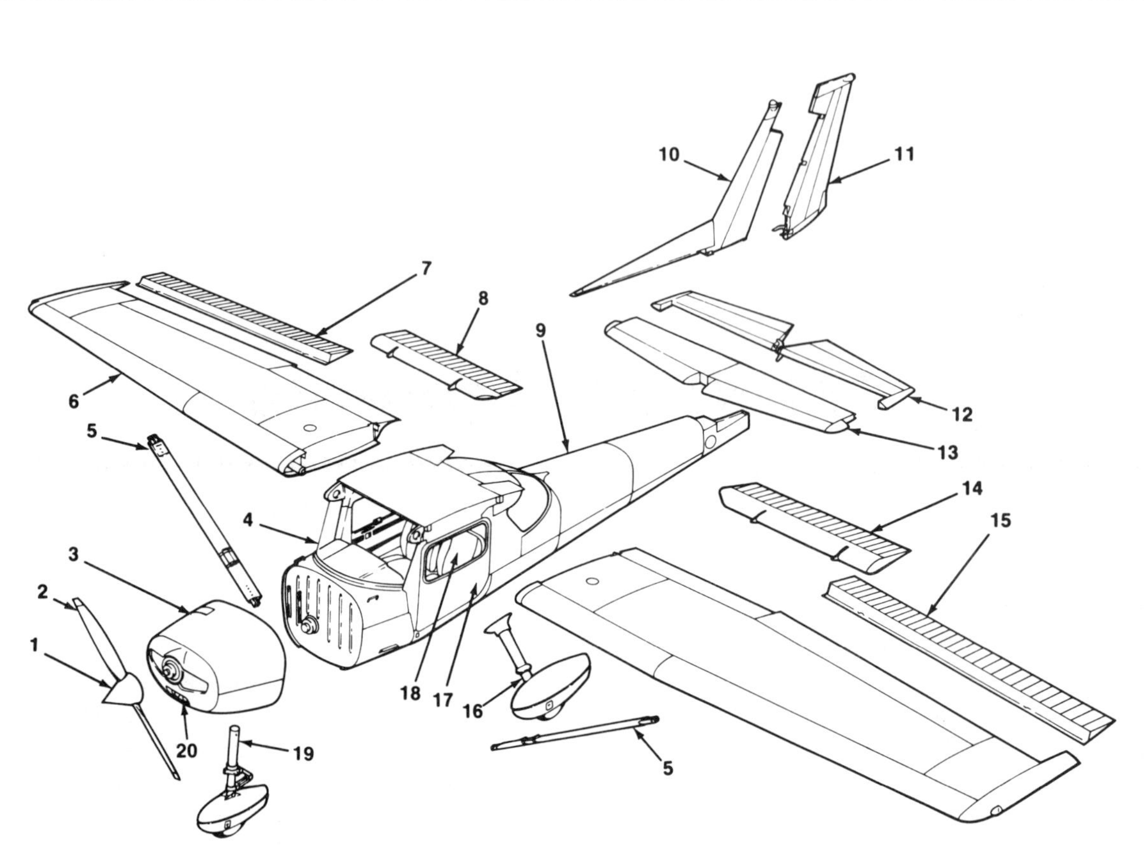

The Main Parts of an Airplane

1. Spinner
2. Propeller
3. Engine Cowl
4. Windshield
5. Wing Strut
6. Wing
7. Right Aileron
8. Right Flap
9. Fuselage
10. Vertical Stabilizer
11. Rudder
12. Elevator
13. Horizontal Stabilizer
14. Left Flap
15. Left Aileron
16. Main Landing Gear
17. Door
18. Seat
19. Nose Gear
20. Landing Lights

(Courtesy Cessna Air Age Education Division)

Acknowledgments

I wish to thank Mr. Fred Bailey, former Chief Ground Instructor, Emery School of Aviation, for his valuable work in making this book technically accurate. Mr. Bailey, along with John Pabst and Barbara Saunders, used this book in their ground school classes, and with the help of many students, were able to make the changes necessary to improve accuracy and comprehension.

To the Emery Staff members including my good friends Rick Tarwater, John Dittmer, Henrietta Groves, and Ruth Sitzman, I owe a debt of gratitude. They made the task of writing this manual easier by their constant and positive support.

A special word of thanks should be expressed to David Emery and Owen Johnson.

I wish to thank Jeppesen Sanderson and Cessna Aircraft Company for permission to reprint published information and photographs.

This book is dedicated to the memory of my brother, Frank Millspaugh, Jr., a bomber pilot killed in action in World War II, and to Bill Mitchell, my first flight instructor. Frank gave me the love of aviation and Bill Mitchell taught me to fly.

Introduction

As a student pilot, you will be responsible both to retain part of what you learn in ground school, and to keep studying throughout your flying career. A lot of your learning will come *after* you get your license. You are going to have to take the initiative to maintain a level of proficiency on your own.

The manual you are about to read is condensed and it would be advisable for you to supplement your understanding with other recognized manuals. For good advice, ask your ground and flight instructors about the availability of well-written general aviation textbooks. You are going to have to "self-direct" much of your learning, and this means that you will be responsible for using all of the resources available to you. Aviation will demand much of you, and it is up to you to develop an attitude of self-learning from the outset. Although 70 percent is a passing score on the FAA examination, a 70 percent landing, in the real world, is a crash!

Earning your Private Pilot Certificate is difficult but certainly not impossible. One important part of the task is passing a written examination conducted by the FAA. That is what this book is all about. I'm going to try to make your written test material correlate to your flight instruction. A lot of facts will come your way and it is your responsibility to study the material for comprehension. There are tons of reference material on just about every subject covered in this book, so don't be afraid to ask an instructor for resource assistance.

Chapter 1

Aircraft Nomenclature

An airplane in flight is a beautiful sight to behold. Unlike the typical automobile, it is designed by its environment. In that regard, it is like a racing yacht, or a streamlined Grand Prix car. The craft is designed for performance, for it is its performance that enables it to fly. The average airplane will almost stop flying at speeds near the 55 mph mark, our U.S. national highway limit. In most automobiles, 100 mph is quite fast, but it is "trainer-plane" speed in the air. The airplane is operated mainly in the speed range of the race car—150 to 200 mph is doing reasonably well for a race car, but is commonplace for most aircraft. Therefore, as you prepare for training as a pilot, you must shift your thinking away from the street and into the sky; you must prepare your mind and your reflexes to become more like the racing car driver.

You should know the airplane thoroughly. This requires a new language and you must learn it well. A thorough knowledge of the nomenclature of the aircraft is a prerequisite to all other phases of flight training and flying experiences. Once you know what you are working *with*, you can embark upon the task of making it work *for* you. It's a whole new, wonderful world; enjoy it—those of us who fly certainly do!

STRUCTURAL UNITS

This chapter on nomenclature will introduce you to the basic airplane (and associated equipment) terminology. It is not necessary for you to know how to disassemble and reassemble the airplane, but a strong working knowledge of the "name of the pieces" is most important to the study of aviation.

The airplane has several structural units and these are the fuselage, the wings, the empennage, the flight controls, and the landing gear. All of these constitute the *airframe.*

The *fuselage* is the main structural unit of an aircraft. The pilot and passengers (or instructor), instruments, baggage, cargo, and other equipment are all contained in the fuselage. On most single-engine aircraft, the engine is connected to the front of the fuselage and a fireproof barrier, called the *firewall*, protects the occupants.

The *wings* are attached to each side of the fuselage and are the principal source of lift. There are countless wing designs and you should know that every wing is a compromise of elements for a particular type of aircraft.

The next section is the *empennage* or "tail." The empennage consists of the *vertical stabilizer* or *fin*, the *horizontal stabilizer*, the *elevator, rudder* and *trim tabs.* The empennage gives the airplane its directional and longitudinal stability and provides the pilot with a means of control (Fig. 1-1).

PRIMARY FLIGHT CONTROLS

The airplane is controllable around its three axes by a deflection of the flight control surfaces, which are hinged against fixed surfaces. Aircraft *attitude*—i.e., *roll, yaw,* and *pitch* (Fig. 1-2)—is controlled by movement of the control surfaces causing aerodynamic action against the airflow. The pilot moves these surfaces by control devices within the cockpit. The elevators and ailerons are controlled by the yoke while the rudder is moved by the rudder pedals (Fig. 1-3).

The *rudder* is attached to the vertical stabilizer (sometimes called the fin) and pro-

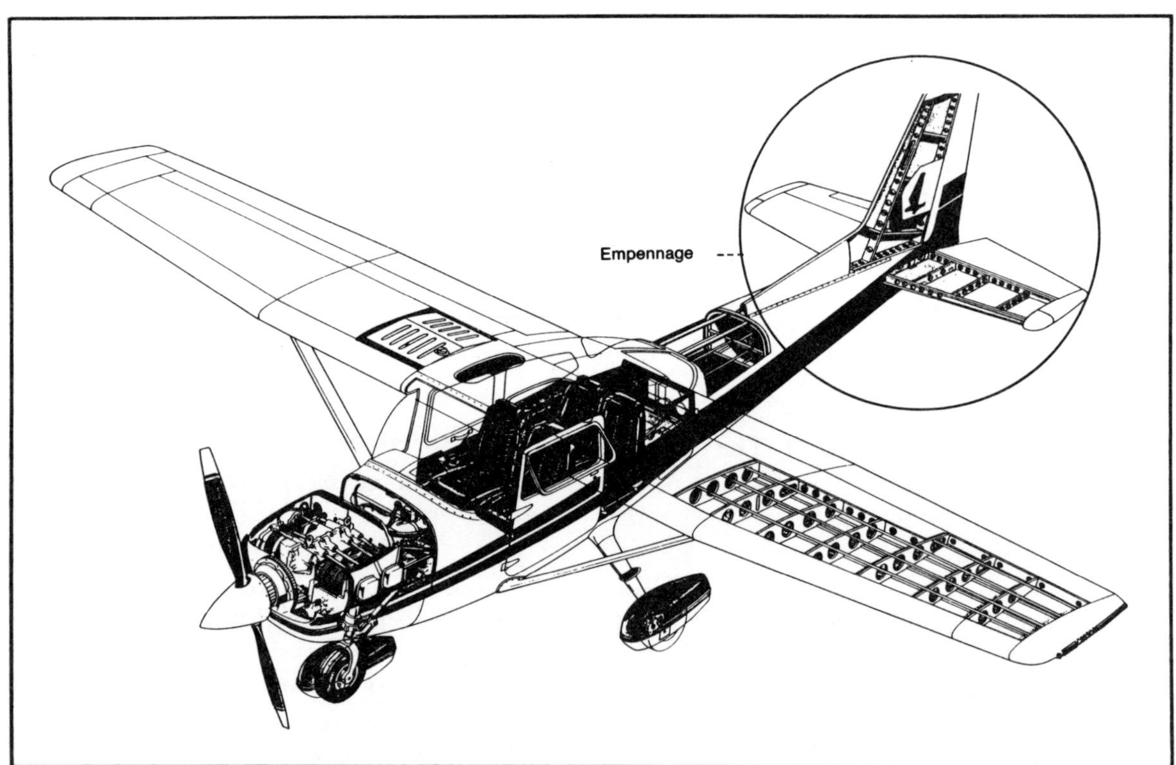

Fig. 1-1. Cessna Skyhawk cutaway. (courtesy Cessna Aircraft Co.)

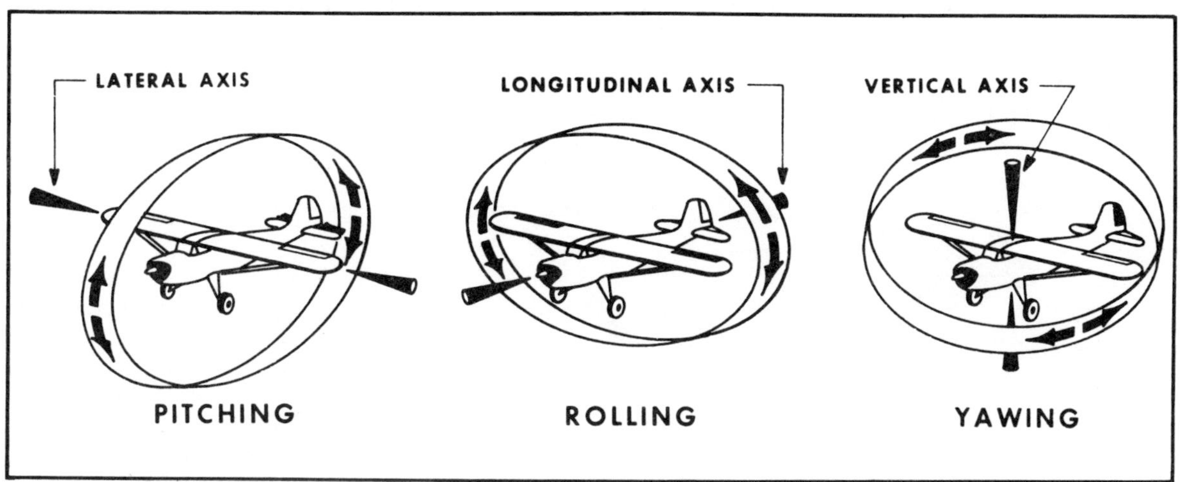

Fig. 1-2. The three axes.

vides a counterforce to an undesirable yawing effect of the ailerons. (This effect is known as *adverse aileron yaw.*) Figure 1-4 will show you how the rudder controls the left or right direction of yaw about the vertical axis. Control of the rudder is by movement of the pedals. When one rudder pedal is pushed forward, the other moves aft, or toward you.

When the rudder is deflected to one side, it extends into the airflow, causing a horizontal force to be exerted in the opposite direction. The rudder is also connected to the nosewheel and action of the rudder pedals provides ground steering. When taxiing an airplane with a tailwheel (a "taildragger"), the prop blast hitting the rudder provides a steering effect that forces the aircraft's nose to move in the desired direction upon deflection.

At the outboard portion of each wing is a movable device called an *aileron*. The ailerons are located on the trailing edge of the wing and, when deflected either up or down, change the wing's surface curvature and its lift. Their primary use is to roll the airplane around the axis that extends through the center from nose to tail, or *longitudinal* axis (Fig. 1-5).

The action of the ailerons causes the aircraft to bank either right or left; the aircraft will move in the direction of the bank.

The ailerons are so connected that they move in opposite directions to each other. As the aileron on one wing moves downward, the other moves upward. This action causes the aircraft to *roll*. To effect a bank, the ailerons are only used momentarily to establish the bank angle and rate of turn that the pilot wants, and then are returned to neutral (or nearly so) until it is time to stop turning or "roll out of the turn." Every pilot has learned that the "hard part" is properly coordinating the rudder and aileron controls to produce a good quality turn.

SECONDARY FLIGHT CONTROLS

The flight controls are broken down into two main divisions, referred to as *primary* and *secondary*. The primary controls comprise the ailerons, elevator, and rudder, while the secondary flight controls consist of the trim tabs and the wing flaps.

The *trim tabs*, as shown in Fig. 1-6, are most commonly used to relieve the pressure on the primary controls. This pressure comes

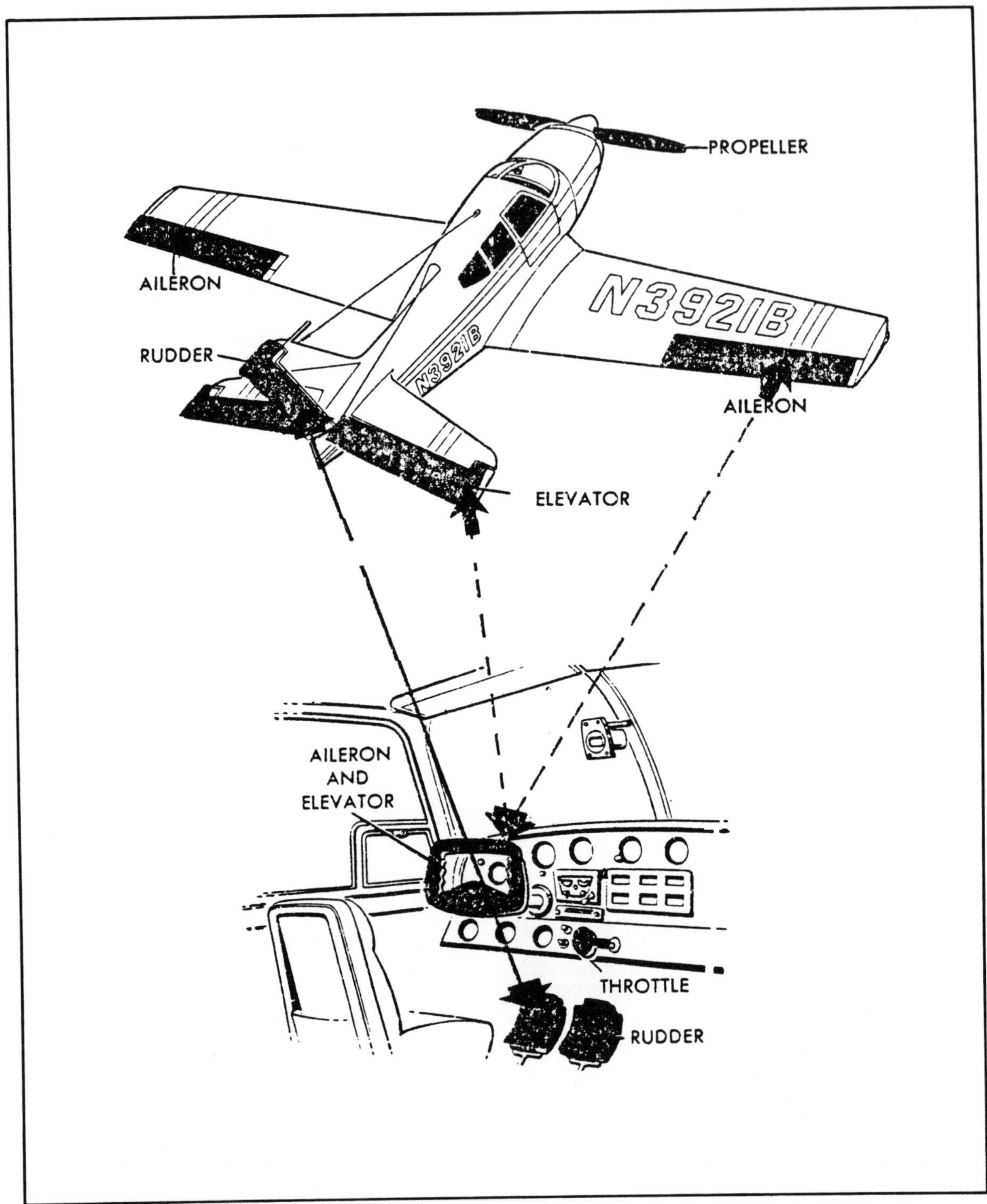

Fig. 1-3. Movable control surfaces and their cockpit controls.

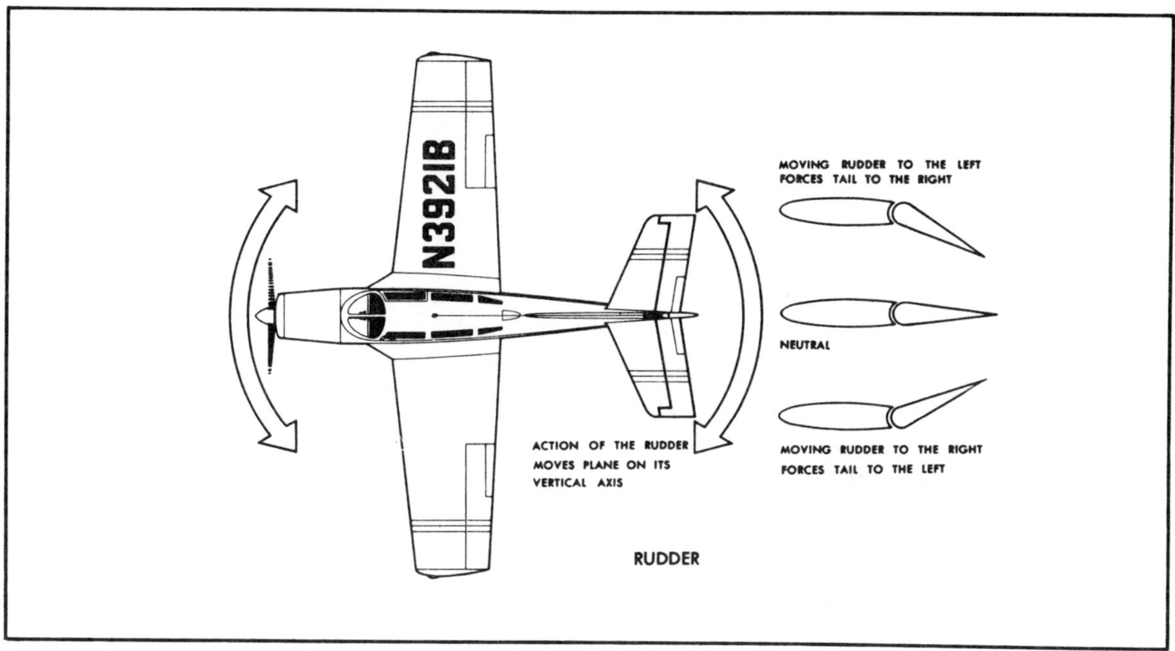

Fig. 1-4. The rudder and its effect.

from the aerodynamic forces imposed by the slipstream (passing air) and other weights. An old-timer once said that "Trim tabs fly more airplanes than pilots!" He was, to be candid, quite right! Trim devices help actuate the main control surfaces by exerting force on those surfaces, thus reducing the amount of force that the a pilot must exert on the controls to maneuver the airplane.

A trim tab is mounted to the primary control surface and it provides a deflection to the airstream that will provide its own up or down force on the surface. Most aircraft have trim devices that are controlled from the cockpit however, some trim devices are adjustable on the ground. The Cessna 152, for example, has a ground-adjustable trim device on the rudder. You might have an instructor show this to you on your next dual session.

Wing flaps are located on the trailing edge of a wing and inboard (close to the fuselage) The main function of wing flaps is to permit a steeper angle of descent. They also permit a slower landing speed and, on many aircraft, may also be used to shorten the takeoff run and provide a steeper climb path (Fig. 1-7).

Flaps are operated in one of three ways: manually, electrically, or hydraulically. When a flap is flush with the wing, we refer to this as having been *retracted.* When a flap is down, we say it is *extended.* Usually the maximum extension is around 30 to 40 degrees. The extended flap increases the lift and drag to that the airplane can descend or climb at a steeper angle which, in effect, means a slower airspeed.

THE OWNER'S MANUAL

The experienced pilot should be familiar with the owner's manual of the aircraft he is planning to fly. In time you will be able to read and understand the entire manual, but for now you can brief yourself by looking up a few key facts in the C-152 owner's manual. For prac-

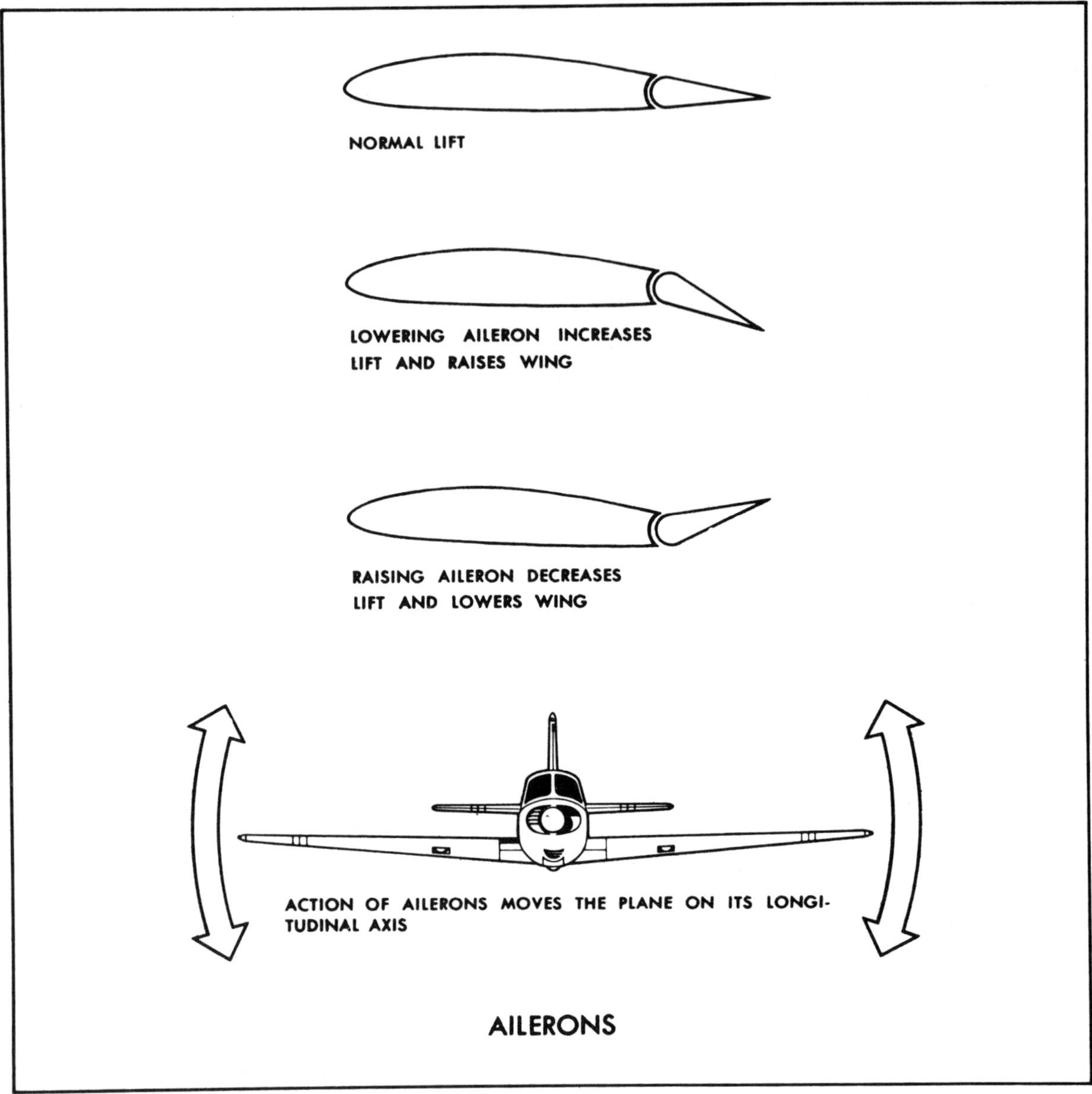

Fig. 1-5. Ailerons.

tice, look up and answer these questions.

Section 1, General

1. *Service ceiling* is the highest altitude at which the aircraft can attain a 100 foot-per-minute climb, under standardized conditions. For the C-152, that is __________ MSL (Mean Sea Level).

2. What is the *usable* fuel capacity with standard tanks?

3. What is the difference between KIAS, KCAS, and KTAS?

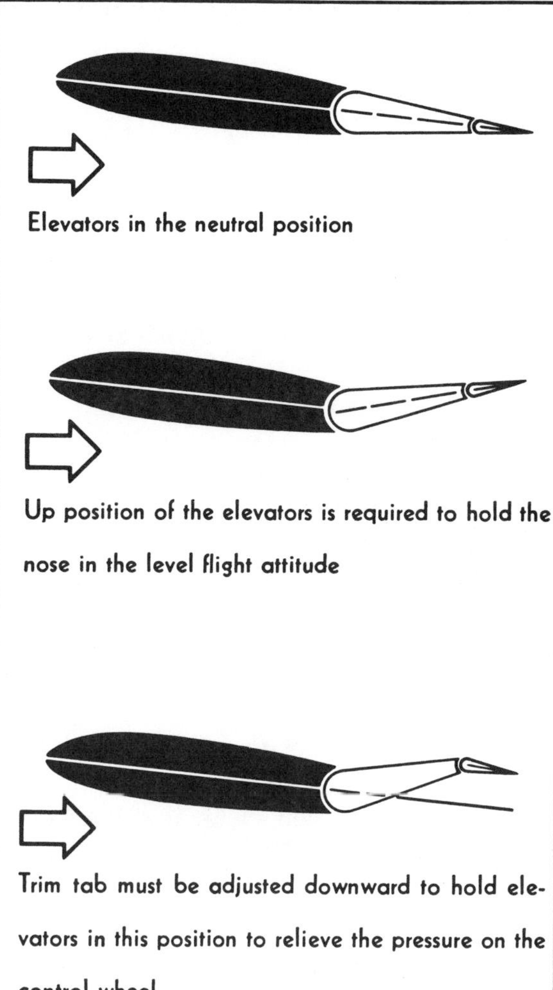

Fig. 1-6. Trim tabs.

Section 2, Limitations

1. Must the aircraft be operated within its Operating Limitations to comply with FAA rules?

2. List the names and numbers for the following V-speeds:

A. V_a
B. V_{ne}
C. V_{fe}

3. Identify the following speed ranges and color codes:

A. 149 KIAS
B. 111 to 149 KIAS
C. 40 to 111 KIAS
D. 35 to 85 KIAS

4. The engine is rated at _________ horsepower at _________ rpm.

5. What is the minimum oil pressure?

6. What is the grade and color of fuel for the C-152?

7. Can you do an intentional spin with the flaps extended?

8. How many quarts will the oil sump hold?

9. Is the C-152 approved for flight into icing conditions?

Section 3, Emergency Procedures

1. In case of engine failure, you should establish the Best Glide Speed, which is _________ KIAS.

2. If an inflight engine fire occurs, you should close the cabin heat control. Why?

3. If an electrical fire exists in flight, what is the *first* thing you should do?

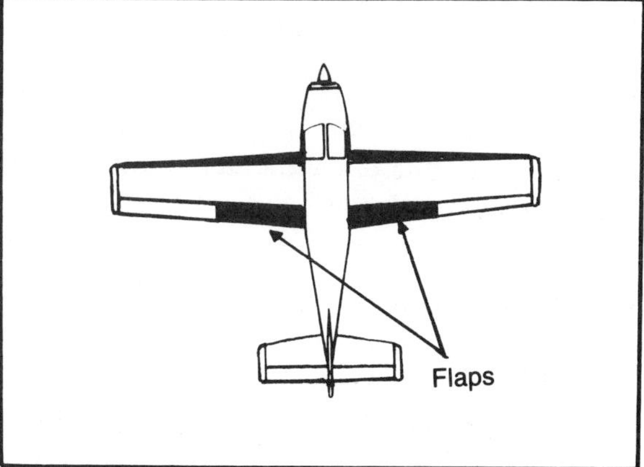

Fig. 1-7. Flaps.

4. At 5,000 feet AGL (Above Ground Level), what is your gliding distance?

5. What are the indications of carburetor icing?

6. In case of a loss of oil pressure during flight, what actions will you take?

Section 4, Normal Procedures

1. What is the maximum allowable "mag drop" for either magneto?

2. What is the maximum difference between the magnetos?

3. Sketch a diagram of the control positions when taxiing with the wind from all positions.

Section 5, Performance

1. At what speed (flaps up) do the IAS and CAS equal each other?

2. With no bank, forward CG, and flaps up, the stall speed is ________ KIAS (V_{s1}).

3. With no bank and 30 degrees of flaps, the stall speed is __________ KIAS with forward CG and __________ KIAS with aft CG. (This is known as V_{so}.)

4. What is the V_y speed for the C-152 and when will you use it?

Section 6, Weight and Balance

1. How much does a gallon of oil weigh?

2. How much does a gallon of fuel weigh?

3. How much weight can you put in the baggage compartment?

4. Maximum Gross Weight for the C-152 is ________ pounds.

Section 7, Systems

1. The maximum nosegear turning angle is __________ degrees.

2. Sketch a diagram of the fuel system.

3. Describe briefly the C-152 electrical system.

4. What instruments are included in the pitot-static system?

5. What instruments are driven by the vacuum system?

6. How soon before the actual stall does the stall horn sound?

Section 8, Service and Handling

1. What one document must be *displayed* in the aircraft at all times?

2. What five documents must be *carried* in the aircraft at all times?

3. What two documents must be made available upon request?

4. How will you clean the windshield and/or windows?

Now that you've passed that hurdle, let's move into the arena of the professional. One of the most important "pro habits" that you can learn is to use a checklist for just about everything. We have the pre-start checklist, the post-start checklist, the pre-takeoff, checklist, the cruise checklist, the landing checklist, etc. To be professional, *do it by the book*!

Chapter 2

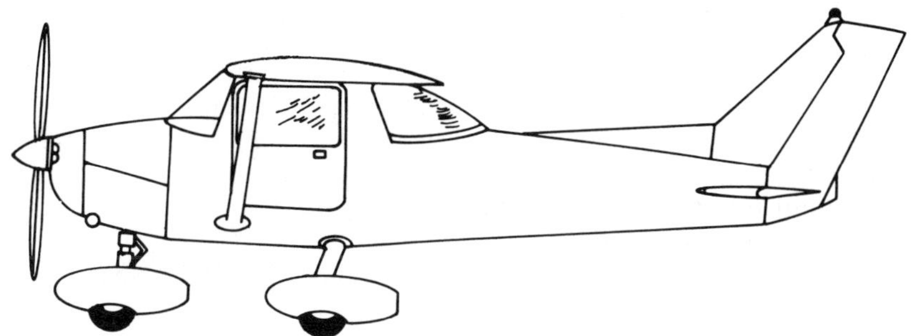

Pre-Solo Preparation

"You gotta do it by the book." Nowhere is the old saying more true than in aviation. The "book" is a set of rules and guidelines governing the operation of an aircraft. The rules are broken down into parts, known as FARs, or *Federal Aviation Regulations.* The parts of concern to us are:

- ☐ FAR Part 1—Official definitions of aviation terms.
- ☐ FAR Part 61—Airman Certification (Pilot and Instructors).
- ☐ FAR Part 91—General Operating Rules.
- ☐ NTSB Part 830—Accident Reporting—as regulated by the National Transportation Safety Board.
- ☐ FAA *Airman's Information Manual*—The non-regulatory "bible" of good operating procedures.
- ☐ Advisory Circulars—Advice from the FAA.

The rules say you must have two things with you at all times: your Pilot Certificate and your current Medical Certificate. Another document you should carry (and we won't mention this one again) is the FCC Radiotelephone Operator's Permit. To get this permit, ask your ground instructor for the application (see Fig. 2-1 for example). Fill out the necessary information. Any further questions—and the application itself—should be directed to: Federal Communications Commission, P.O. Box 1050, Gettysburg, PA 17325. Notice the form is divided into Parts I, II, and III—you should keep Part III in your possession as a temporary permit. Mail Parts I and II to the above address, and the FCC should return a permanent Operator's Permit within 60 days.

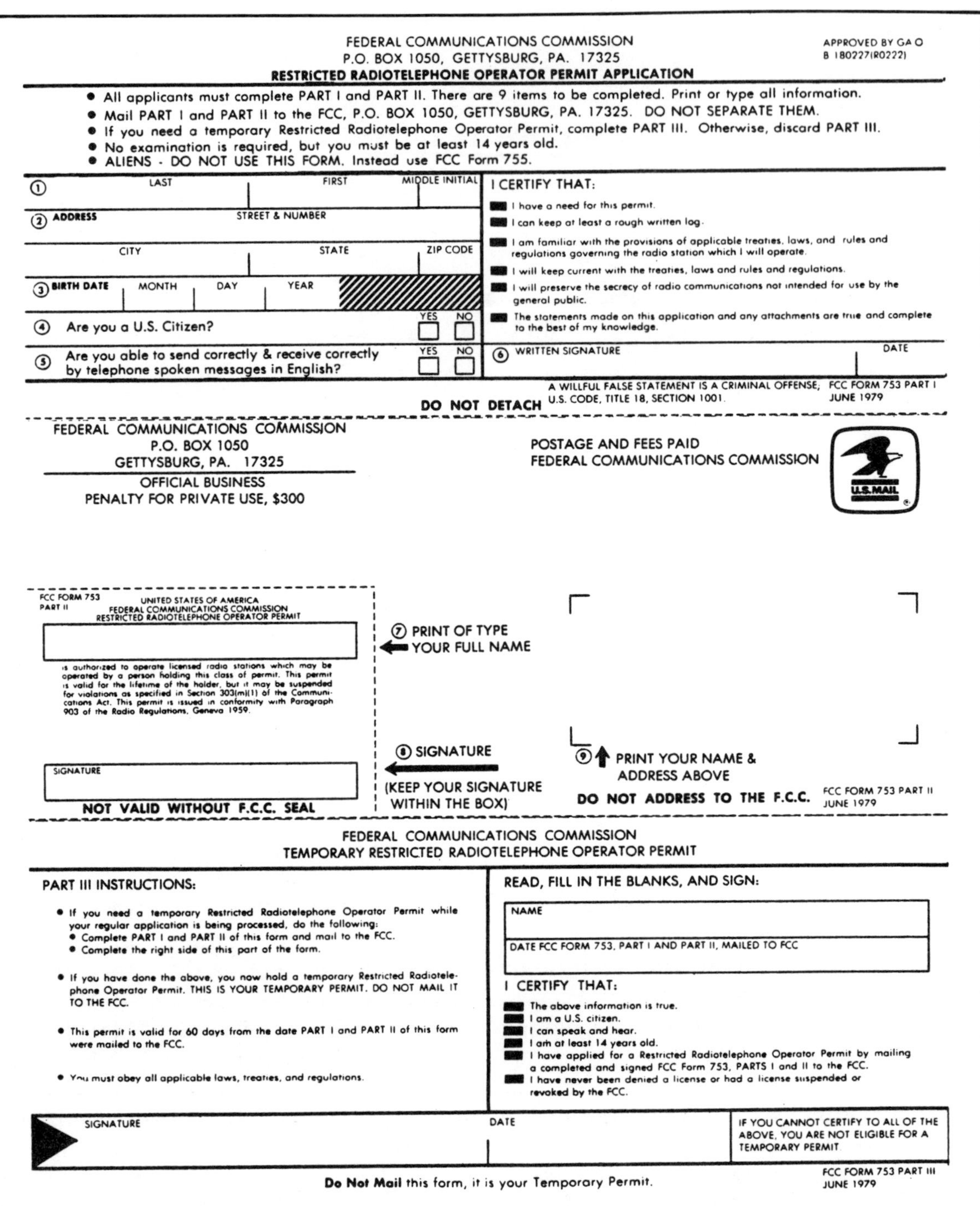

FEDERAL COMMUNICATIONS COMMISSION
P.O. BOX 1050, GETTYSBURG, PA. 17325
RESTRICTED RADIOTELEPHONE OPERATOR PERMIT APPLICATION

APPROVED BY GAO
B 180227(R0222)

- All applicants must complete PART I and PART II. There are 9 items to be completed. Print or type all information.
- Mail PART I and PART II to the FCC, P.O. BOX 1050, GETTYSBURG, PA. 17325. DO NOT SEPARATE THEM.
- If you need a temporary Restricted Radiotelephone Operator Permit, complete PART III. Otherwise, discard PART III.
- No examination is required, but you must be at least 14 years old.
- ALIENS - DO NOT USE THIS FORM. Instead use FCC Form 755.

① LAST | FIRST | MIDDLE INITIAL

② ADDRESS | STREET & NUMBER

CITY | STATE | ZIP CODE

③ BIRTH DATE | MONTH | DAY | YEAR

④ Are you a U.S. Citizen? YES ☐ NO ☐

⑤ Are you able to send correctly & receive correctly by telephone spoken messages in English? YES ☐ NO ☐

I CERTIFY THAT:

- I have a need for this permit.
- I can keep at least a rough written log.
- I am familiar with the provisions of applicable treaties, laws, and rules and regulations governing the radio station which I will operate.
- I will keep current with the treaties, laws and rules and regulations.
- I will preserve the secrecy of radio communications not intended for use by the general public.
- The statements made on this application and any attachments are true and complete to the best of my knowledge.

⑥ WRITTEN SIGNATURE | DATE

A WILLFUL FALSE STATEMENT IS A CRIMINAL OFFENSE; U.S. CODE, TITLE 18, SECTION 1001.

FCC FORM 753 PART I
JUNE 1979

DO NOT DETACH

FEDERAL COMMUNICATIONS COMMISSION
P.O. BOX 1050
GETTYSBURG, PA. 17325

OFFICIAL BUSINESS
PENALTY FOR PRIVATE USE, $300

POSTAGE AND FEES PAID
FEDERAL COMMUNICATIONS COMMISSION

U.S. MAIL

FCC FORM 753
PART II

UNITED STATES OF AMERICA
FEDERAL COMMUNICATIONS COMMISSION
RESTRICTED RADIOTELEPHONE OPERATOR PERMIT

is authorized to operate licensed radio stations which may be operated by a person holding this class of permit. This permit is valid for the lifetime of the holder, but it may be suspended for violations as specified in Section 303(m)(1) of the Communications Act. This permit is issued in conformity with Paragraph 903 of the Radio Regulations, Geneva 1959.

SIGNATURE

NOT VALID WITHOUT F.C.C. SEAL

⑦ PRINT OF TYPE YOUR FULL NAME

⑧ SIGNATURE

(KEEP YOUR SIGNATURE WITHIN THE BOX)

⑨ PRINT YOUR NAME & ADDRESS ABOVE

DO NOT ADDRESS TO THE F.C.C.

FCC FORM 753 PART II
JUNE 1979

FEDERAL COMMUNICATIONS COMMISSION
TEMPORARY RESTRICTED RADIOTELEPHONE OPERATOR PERMIT

PART III INSTRUCTIONS:

- If you need a temporary Restricted Radiotelephone Operator Permit while your regular application is being processed, do the following:
  - Complete PART I and PART II of this form and mail to the FCC.
  - Complete the right side of this part of the form.
- If you have done the above, you now hold a temporary Restricted Radiotelephone Operator Permit. THIS IS YOUR TEMPORARY PERMIT. DO NOT MAIL IT TO THE FCC.
- This permit is valid for 60 days from the date PART I and PART II of this form were mailed to the FCC.
- You must obey all applicable laws, treaties, and regulations.

READ, FILL IN THE BLANKS, AND SIGN:

NAME

DATE FCC FORM 753, PART I AND PART II, MAILED TO FCC

I CERTIFY THAT:

- The above information is true.
- I am a U.S. citizen.
- I can speak and hear.
- I am at least 14 years old.
- I have applied for a Restricted Radiotelephone Operator Permit by mailing a completed and signed FCC Form 753, PARTS I and II to the FCC.
- I have never been denied a license or had a license suspended or revoked by the FCC.

SIGNATURE | DATE | IF YOU CANNOT CERTIFY TO ALL OF THE ABOVE, YOU ARE NOT ELIGIBLE FOR A TEMPORARY PERMIT.

Do Not Mail this form, it is your Temporary Permit.

FCC FORM 753 PART III
JUNE 1979

Fig. 2-1. FCC Radio Operator Permit.

YOUR MEDICAL

The pilot-to-be must get his health checked by an FAA Medical Doctor. Your ground instructor will have a list of local physicians that are approved by the FAA. The medical examination is not difficult and I have found it to be quite comparable to a simple insurance physical. There are three classes of Medical Certificates:

- ☐ First Class—Required for Airline Pilot operations.
- ☐ Second Class—Required for Commercial Pilot operations.
- ☐ Third Class—Required for Student and Private Pilot operations.

Some pilots want to continue their flight training right through to the professional level. If you plan to fly commercially, you may choose to get the First Class or Second Class Medical instead of the "minimum" Third Class. All medical certificates are valid for Student or Private privileges for a term of 24 calendar months. ("Calendar" means to the end of the month.) The Second Class is valid for one year of Commercial privileges, followed by another year limited to Student or Private privileges. Likewise, the First Class is valid for six months of Airline work or one year of Commercial flying, but the usual full two years of Student/Private. All medicals expire at the end of the month indicated.

Figure 2-2 is an example of the typical "combined" Third Class Medical and Student Pilot Certificate. Remember that it must be in your personal possession at all times when exercising your flight privileges.

Notice that the back of the Student Pilot Certificate includes spaces for your instructor's signature as he "signs you off" for solo, and again for solo cross-country flying. Solo flights are not legal without his endorsement.

YOUR "TICKET"

Upon completion of your Private Pilot training, having passed the written, oral, and flight tests, you will be issued a temporary license, shown in Fig. 2-3. This will be your Private Certificate until the permanent copy, also shown, arrives from the FAA's Oklahoma City office.

On the certificates shown in Figs. 2-3 and 2-4, note the words "airplane single engine

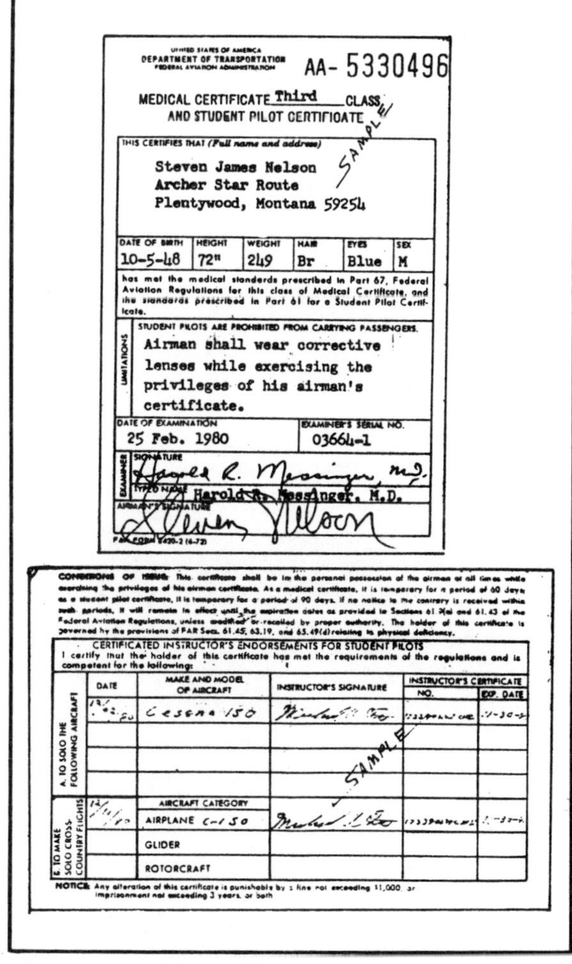

UNITED STATES OF AMERICA
DEPARTMENT OF TRANSPORTATION
FEDERAL AVIATION ADMINISTRATION

AA-5330496

MEDICAL CERTIFICATE Third CLASS
AND STUDENT PILOT CERTIFICATE

SAMPLE

THIS CERTIFIES THAT *(Full name and address)*

Steven James Nelson
Archer Star Route
Plentywood, Montana 59254

DATE OF BIRTH	HEIGHT	WEIGHT	HAIR	EYES	SEX
10-5-48	72"	249	Br	Blue	M

has met the medical standards prescribed in Part 67, Federal Aviation Regulations for this class of Medical Certificate, and the standards prescribed in Part 61 for a Student Pilot Certificate.

LIMITATIONS: STUDENT PILOTS ARE PROHIBITED FROM CARRYING PASSENGERS.
Airman shall wear corrective lenses while exercising the privileges of his airman's certificate.

DATE OF EXAMINATION	EXAMINER'S SERIAL NO.
25 Feb. 1980	03664-1

EXAMINER: SIGNATURE Harold R. Messinger, M.D.
TYPED NAME Harold R. Messinger, M.D.

AIRMAN'S SIGNATURE Steven Nelson

FAA FORM 8420-2 (6-72)

CONDITIONS OF ISSUE: This certificate shall be in the personal possession of the airman at all times while exercising the privileges of his airman certificate. As a medical certificate, it is temporary for a period of 60 days; as a student pilot certificate, it is temporary for a period of 90 days. If no notice to the contrary is received within such periods, it will remain in effect until the expiration dates as provided in Sections 61.23(a) and 61.43 of the Federal Aviation Regulations, unless modified or recalled by proper authority. The holder of this certificate is governed by the provisions of FAR Secs. 61.45, 63.19, and 65.49(d) relating to physical deficiency.

CERTIFICATED INSTRUCTOR'S ENDORSEMENTS FOR STUDENT PILOTS

I certify that the holder of this certificate has met the requirements of the regulations and is competent for the following:

	DATE	MAKE AND MODEL OF AIRCRAFT	INSTRUCTOR'S SIGNATURE	INSTRUCTOR'S CERTIFICATE NO.	EXP. DATE
A. TO SOLO THE FOLLOWING AIRCRAFT	[illegible]	Cessna 150	[illegible]	[illegible]	[illegible]
B. TO MAKE SOLO CROSS-COUNTRY FLIGHTS		AIRCRAFT CATEGORY			
	[illegible]	AIRPLANE C-150	[illegible]	[illegible]	[illegible]
		GLIDER			
		ROTORCRAFT			

NOTICE: Any alteration of this certificate is punishable by a fine not exceeding $1,000, or imprisonment not exceeding 3 years, or both

Fig. 2-2. Medical and Student Certificate.

I. UNITED STATES OF AMERICA
DEPARTMENT OF TRANSPORTATION – FEDERAL AVIATION ADMINISTRATION

ii. **TEMPORARY AIRMAN CERTIFICATE**

III. CERTIFICATE NO. 523467864

THIS CERTIFIES THAT IV. JOHN A. DOE
V. 224 W. LAMAR ST.
HOLLIS, OKLAHOMA 73550

DATE OF BIRTH	HEIGHT	WEIGHT	HAIR	EYES	SEX	NATIONALITY VI.
02-13-39	67 IN.	160	BROWN	BROWN	M	USA

IX. **has been found to be properly qualified and is hereby authorized in accordance with the conditions of issuance on the reverse of this certificate to exercise the privileges of**

PRIVATE PILOT

RATINGS AND LIMITATIONS

XII. AIRPLANE SINGLE ENGINE LAND

XIII.

THIS IS ☒ AN ORIGINAL ISSUANCE ☐ A REISSUANCE OF THIS GRADE OF CERTIFICATE

DATE OF SUPERSEDED AIRMAN CERTIFICATE

BY DIRECTION OF THE ADMINISTRATOR

EXAMINER'S DESIGNATION NO. OR INSPECTOR'S REG. NO. RM 03-27

X. DATE OF ISSUANCE 05-15-80

X. SIGNATURE OF EXAMINER OR INSPECTOR JOHN A. DITTMER

DATE DESIGNATION EXPIRES 12-31-81

vii. AIRMAN'S SIGNATURE

FAA Form 8060-4 (8-79) USE PREVIOUS EDITION

Fig. 2-3. The temporary Airman Certificate.

land." This represents the major limitation upon the pilot's privileges, as he is not authorized to fly any *category* of aircraft except "airplane," nor is he legal for any *class* of airplane except "single engine land." In other words, the category of "glider" would require an additional rating to be added to the certificate, as would the class of "single engine sea" or "multi-engine land." So the words *category* and *class* are used to define the pilot's authority.

THE ALL-IMPORTANT CHECKLIST

One of the first steps toward actually flying an airplane is called the *preflight inspection*.

I. UNITED STATES OF AMERICA XI.
Department of Transportation – FEDERAL AVIATION ADMINISTRATION
THIS CERTIFIES THAT IV.
V. LEISTER BRYANT STOTTLEMYER JR
RT 1 BOX 449
SMITHSBURG MD 21783

DATE OF BIRTH	HEIGHT IN.	WEIGHT	HAIR	EYES	SEX	NATIONALITY VI.
05-22-30	70	225	BROWN	BLUE	M	USA

IX. HAS BEEN FOUND TO BE PROPERLY QUALIFIED TO EXERCISE THE PRIVILEGES OF
II. PRIVATE PILOT
III. CERT. NO. 2153[illegible]
RATINGS AND LIMITATIONS
XII. AIRPLANE SINGLE ENGINE LAND
XIII.
VII. SIGNATURE OF HOLDER
X. DATE OF ISSUE: 12-29-82
VIII. ADMINISTRATOR
AC Form 8060-2 (5-81)

Fig. 2-4. The permanent Airman Certificate.

It is vitally important that you learn this procedure early and stick with it the rest of your flying career, so let's preflight the bird on paper—you'll have help on the real one.

The control lock on the *yoke* (control wheel) is removed and stored in the glove compartment. Check the reading on the tach sheet and compare this with the Hobbs meter. If it agrees, place the tach sheet in the glovebox to help eliminate clutter. Make sure that all electrical switches are OFF. Place the ignition key on the floor (never in the switch) to ensure that the engine can't start. Turn on the master switch to check the fuel gauges and lower the flaps. Turn the master switch OFF.

Now, start at the pilot's door and methodically work your way around the airplane (Fig. 2-5) to check the following items:

- ☐ All surfaces for dents, popped rivets, and other damage.
- ☐ Horizontal stabilizer attachment points and tips.
- ☐ Left elevator, counterweights, tip, attachment points, and for freedom of movement.
- ☐ Rudder attachment points, count-counterweight, nav light, beacon, control cables and control stops for the elevator and rudder, and untie tail tiedown (no wind).
- ☐ Right elevator attachment points and freedom of movement, trim tab hinge and control rod, elevator tip and attach point.
- ☐ Right horizontal stabilizer tip and attachment point.
- ☐ Right fuselage side for dents, popped rivets, damage signs.
- ☐ Right flap from the rear for freedom from binding of the rollers, security of actuator arm, and, from underneath, security of inspection panels.
- ☐ Right aileron freedom of movement, hinges and cotter keys, actuator arm and bolts, and counterweights.
- ☐ Right wingtip for security and condition, nav light.
- ☐ Leading edge of right wing for general condition, untie wing tiedown (no wind).
- ☐ Right landing gear for proper tire inflation, no cords showing on tire, clearance between the strut and hydraulic brake line, security of brakes, no hydraulic leaks.
- ☐ Right side of cowling for cabin air and instrument panel cooling vents.
- ☐ Engine compartment door for security of hinges and fasteners, oil level, loose wires or oil leaks; take fuel sample.
- ☐ Right side of nosewheel for security of steering arm and seal, proper inflation and condition of nosewheel strut and absence of hydraulic leaks, crankcase breathers, oil quick-drain tube and gas drain tube.

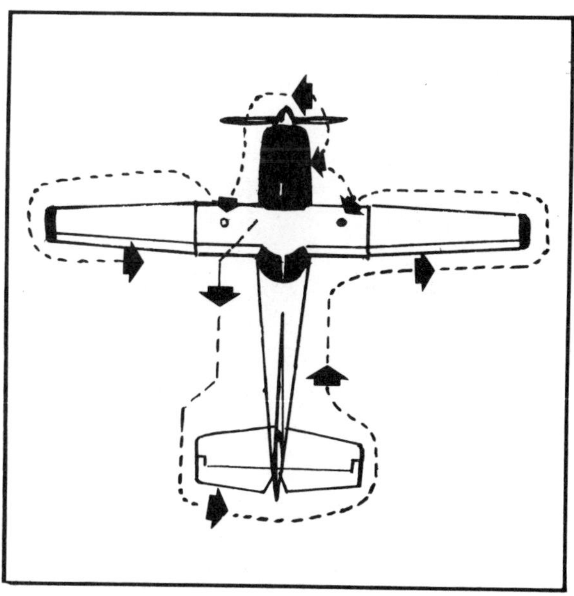

Fig. 2-5. Exterior inspection.

- ☐ Prop for nicks and general condition. Also, absence of foreign material on the cylinders inside cowling, cabin, and carb heat intakes. Inspect spinner for security.
- ☐ Check the left side of the aircraft the same as you did on the right side, except for the static port on the left side of the fuselage, the pitot tube with drain hole, the fuel vent, and the landing light, all on the left wing.
- ☐ Visually check the fuel, filler caps, and secure.

THE INSTRUMENT PANEL

Looking at the instrument panel for the first time is not only fascinating, but sometimes frightening. It will, however, become very familiar to you as you go along. Study the illustration of the panel in Fig. 2-6 and test yourself, then relate this to the actual instrument panel. The sooner you learn the layout, the better off you will be.

Looking at the instrument panel, left to right, top to bottom, you will see the same basic layout in virtually every training aircraft you may fly. The instruments directly in front of the pilot (left seat) are the flight instrument group, which starts in the upper left with the *airspeed indicator.* Next, going right, is the *attitude indicator,* followed by the *clock.* The *altimeter* (number 7) is next. Behind the handgrip of the control yoke is the *turn coordinator.* Just behind the yoke is the heading indicator or "DG" (directional gyro). Number 20 is the *vertical speed indicator,* which completes the flight instruments. Numbers 8 and 21, next to the flight group, are the VOR and ADF, two radio navigation instruments you will learn about later.

Engine gauges are grouped on the right side and/or below the flight instruments. This group includes the *tachometer* which measures engine speed in revolutions per minute (rpm), two gauges for the *temperature* and *pressure* of the engine lubricating oil, *fuel gauge* for each of the two fuel tanks, and an *ammeter* to show whether the electrical system is charging or draining the battery.

THE FLIGHT INSTRUMENTS

To better understand what the flight instruments will tell you, let's take a closer look at that group.

The *airspeed indicator* (Fig. 2-7) is a gauge that works on pressure differential between air that is moving and air that is static or undisturbed. The moving air creates a ram pressure at the pitot tube as the aircraft moves through the air. The static air is the undisturbed atmospheric pressure at the level of flight, as measured at the static port. These two pressures will be equal when the airplane is sitting stationary on the ground in quiet air. When the aircraft flies, the pressure in the pitot is greater than the pressure in the static line. The airspeed indicator has a pointer on the face which is calibrated to read this pressure difference as *miles per hour* or *knots* (nautical miles per hour). The color codes on the face of the dial have a special meaning to the pilot in that they quickly indicate some of the important V-speeds that you've already started to learn.

Looking at the panel, again left to right, the next instrument is the *attitude indicator,* sometimes called the *artificial horizon* (Fig. 2-8). This instrument has a display of a miniature airplane and a bar representing the horizon. The bank attitude is shown by the scale across the upper part of the instrument and graduated in degrees, from 10 to 90 degrees. Pitch attitude is also shown by the distance of the horizon bar above or below the

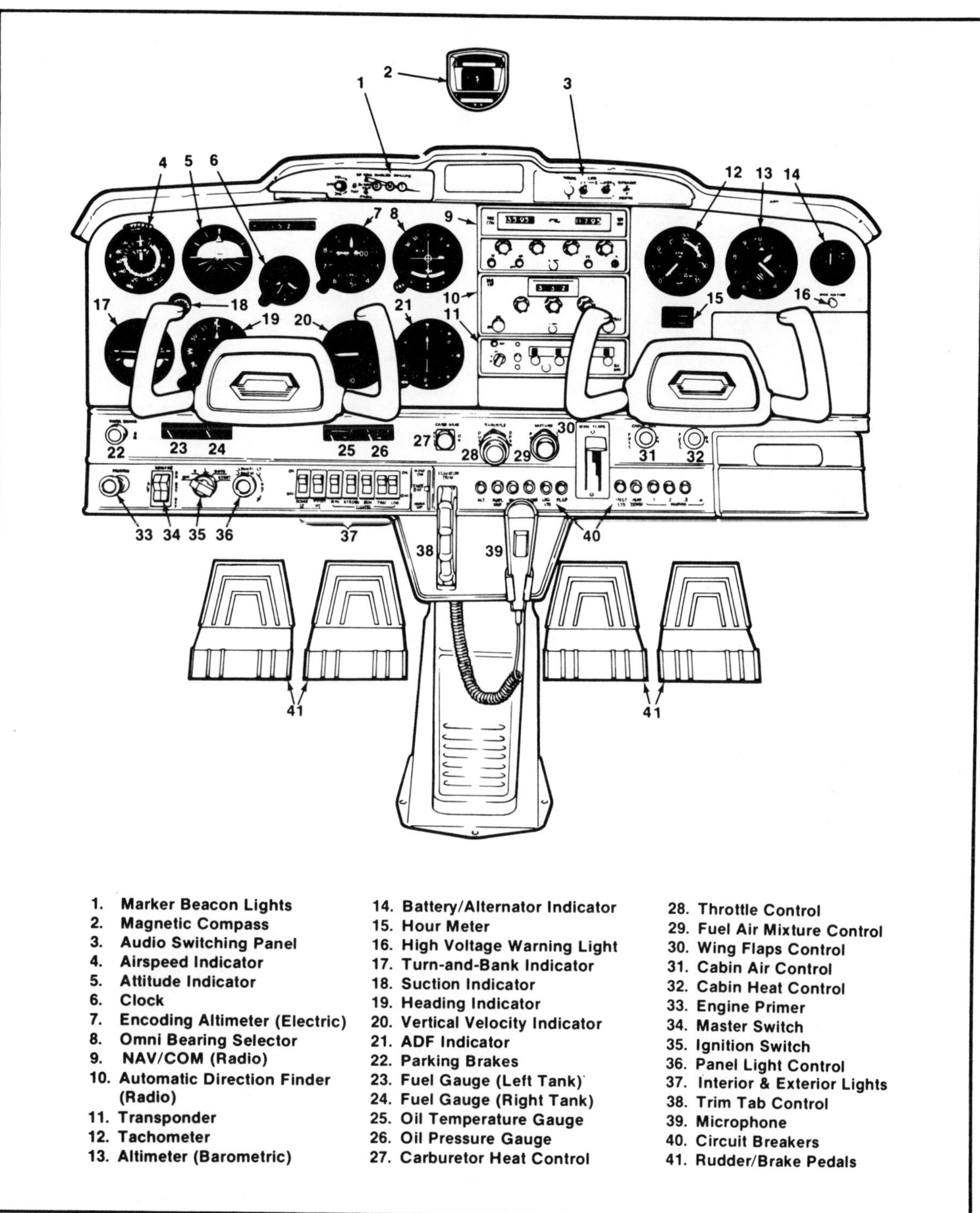

Fig. 2-6. Instrument panel.

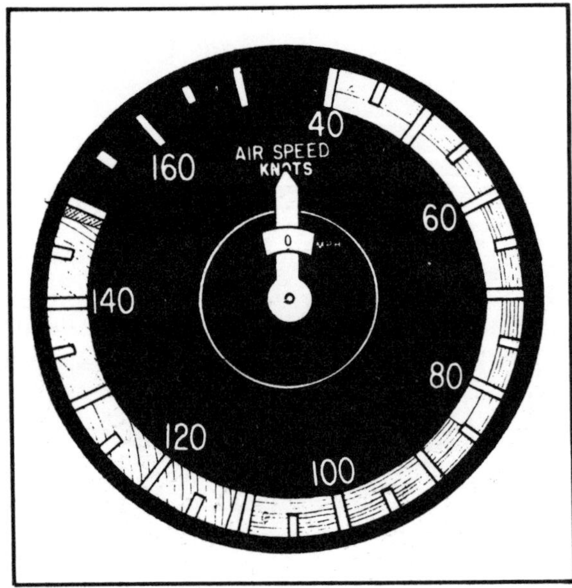

Fig. 2-7. Airspeed indicator.

nose of the symbolic airplane, which is the center dot. So the relationship of the miniature aircraft to the horizon bar is the same as the relationship of the real aircraft to the actual horizon. This instrument gives an instantaneous indication of the smallest changes in attitude (pitch and bank).

Be sure you can interpret the banking scale, since the scale indicators move in the opposite direction from that in which the aircraft is actually banked. The scale marks are for banks of 0, 10, 20, 30, 60, and 90 degrees.

Just to show you how this thing can confuse you, look at Fig. 2-9 and put your hand over the airplane (left of the instrument) and tilt the page until the symbolic airplane in the center is level to you . . . then note the horizon bar . . . that's the way it looks in the real airplane!

The *altimeter* is the only instrument giving us altitude information, so it's one of the important ones. Simply stated, it's a sort of barometer that measures outside air pressure (through the static source) to give a readout of altitude above sea level. The altimeter is adjustable to compensate for variations in pressure caused by changing weather systems. The pilot must periodically reset the "window" of the altimeter to the known sea level pressure, or set the hands to a known field elevation.

Note the large hand. As it starts to move to the right from sea level, where all hands are on 0, each increment or mark is 20 feet. The pointer would pass 20, then 40, then 60 . . . 100, and so on. After making one complete revolution, the mid-size pointer will pass the 1, which represents 1,000 feet, and so on. The altimeter shown in Fig. 2-10 is set to 30.34

Fig. 2-8. Attitude indicator.

inches of mercury and reads 100 feed above sea level.

Although the altimeter in Fig. 2-10 didn't show it, there is a third small needle. It is clearly shown in Fig. 2-11 and moves to the 1 at 10,000 feet.

The *turn coordinator* (Fig. 2-12) or the older style *turn and slip indicator* (Fig. 2-13) shows the direction and rate at which the aircraft is turning, as well as whether or not the aircraft is slipping or skidding sideways.

Each of these is actually two instruments combined—one is a turn needle and the other is a ball placed in a fluid. The needle is connected to a gyro and depends on its gyroscopic principals for its measurement of the rate of turn. The ball is actuated by a combination of gravity and centrifugal force (which we call *load factor* or Gs). If the rudder and ailerons are properly coordinated in flight, the combination of forces or load factor causes the ball to rest in the lowest part of the tube centered between reference marks. When the forces become unbalanced, the ball moves away from the center of the tube.

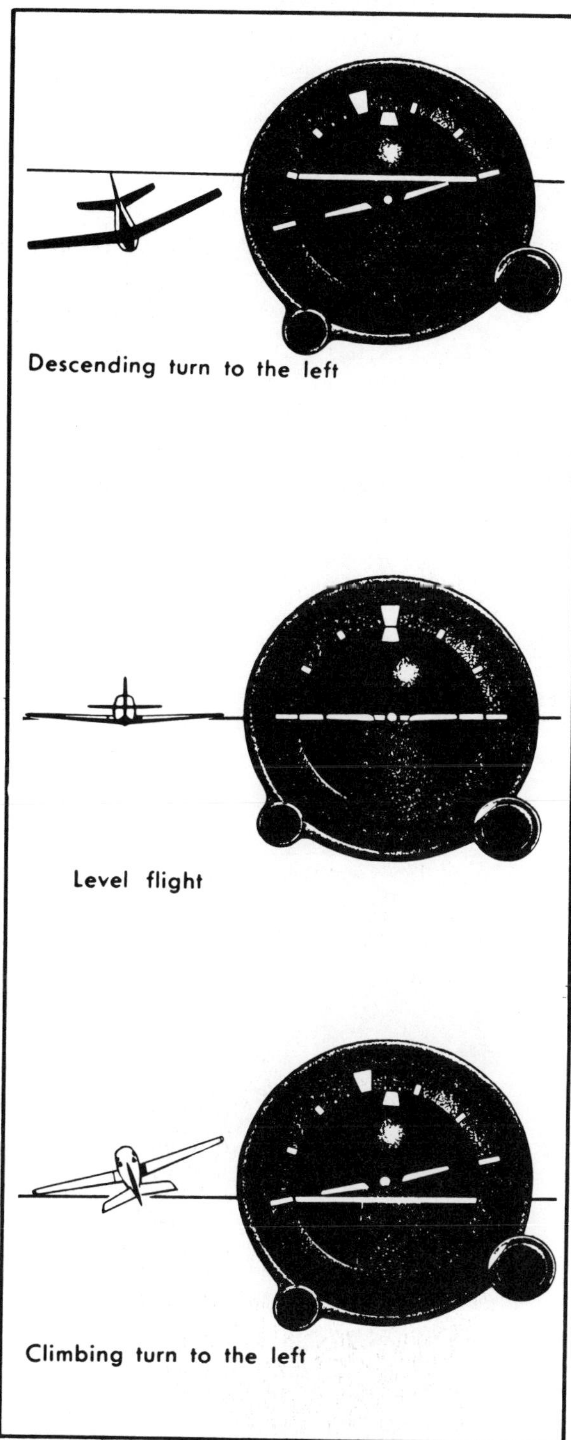

Fig. 2-9. Interpreting the attitude indicator.

Fig. 2-10. Altimeter.

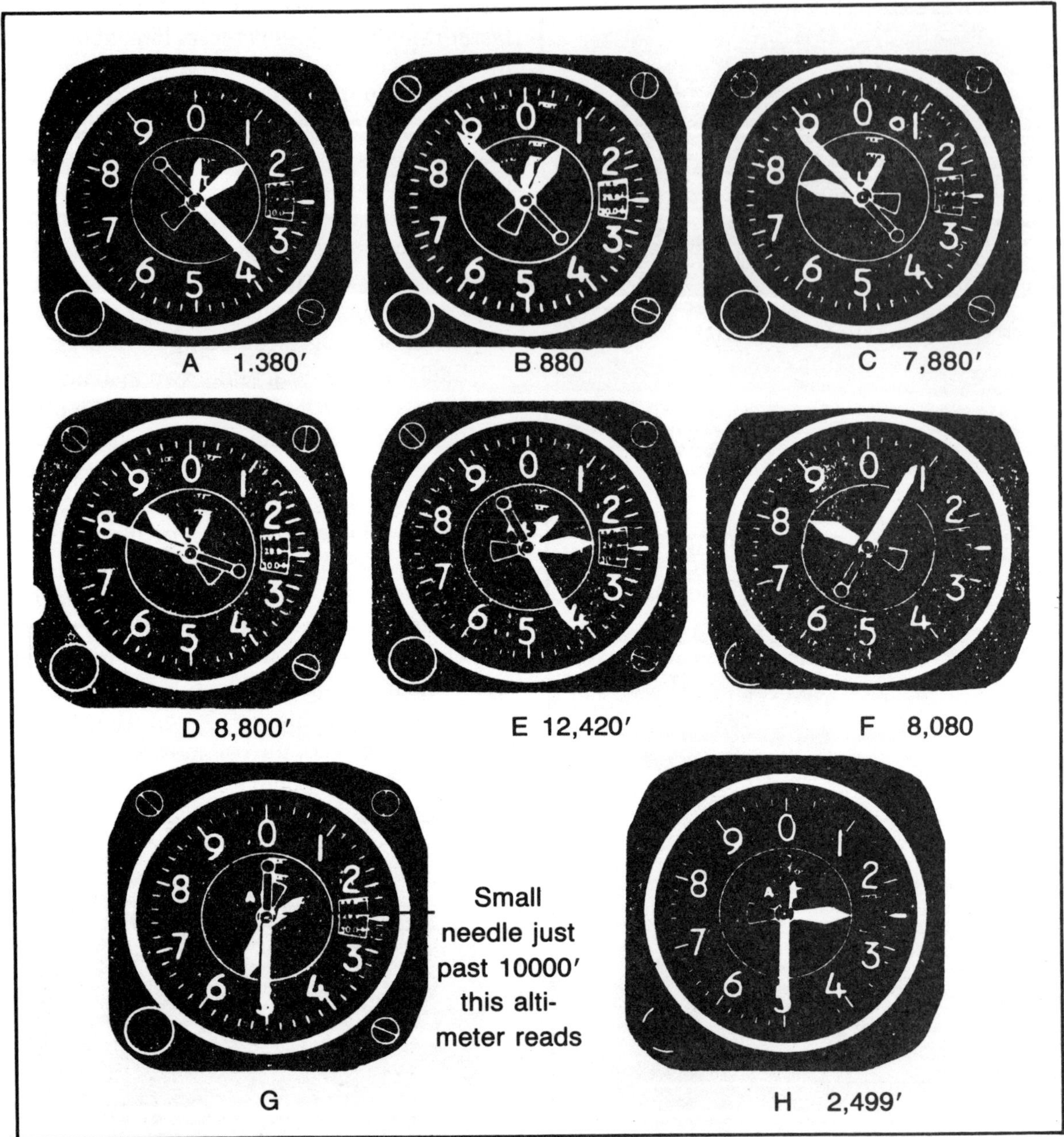

Fig. 2-11. Altimeter interpretations.

The turn coordinator displays a small aircraft on the roll axis that banks proportionally to the rate of turn. If the wings are aligned with the reference marks, the aircraft will turn 360 degrees in exactly two minutes, or three degrees per second. The turn coordinator also has a ball like the turn and slip.

It should be noted that neither of these in-

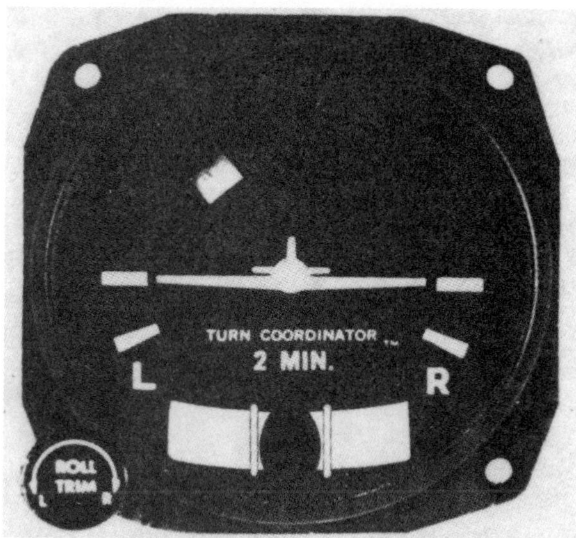

Fig. 2-12. Turn coordinator.

struments gives an actual indication of the degree of bank. For any given airspeed, there is an angle of bank necessary to maintain a "two minute turn" or what we call a *standard rate* turn. The faster the airspeed, the greater the angle of bank necessary to obtain any given rate of turn. A little trainer can turn at a much faster rate than a jet fighter. (Think about that one!)

The DG (*directional gyro*) is a mechanical compass, for all practical purposes (Fig. 2-14). The pilot must set it to agree with the magnetic compass and then reset it at periodic intervals. The DG is gyroscopically stabilized and, as such, is not subject to the turbulence that causes the magnetic compass to read erroneously when flying in any manner except straight and level in smooth air. The DG is, however, subject to drift or "precession" and for this reason it must be periodically reset to maintain an accurate heading.

The scale of its dial is similar to the magnetic compass but easier to read. It has a small airplane outlined pointing to the present compass heading. (330 degrees shown).

The *VSI* (*vertical speed indicator*), sometimes called the "rate of climb," is a form of altimeter which is connected to the static port of the airplane (Fig. 2-15). Inside the case

Fig. 2-13. Turn and slip indicator.

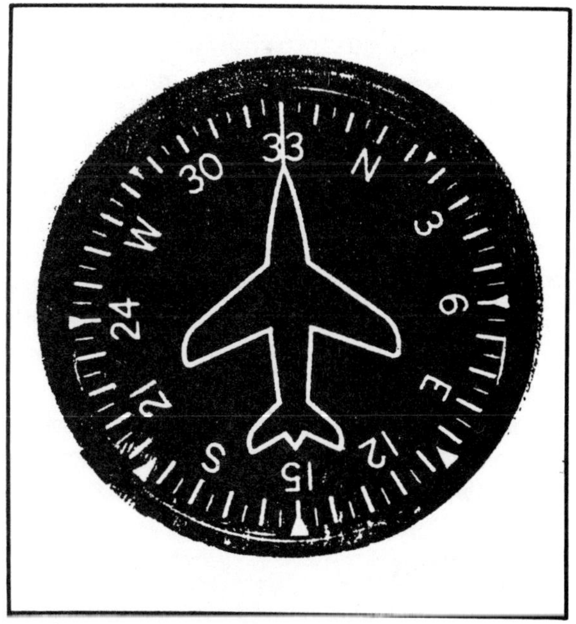

Fig. 2-14. Directional gyro.

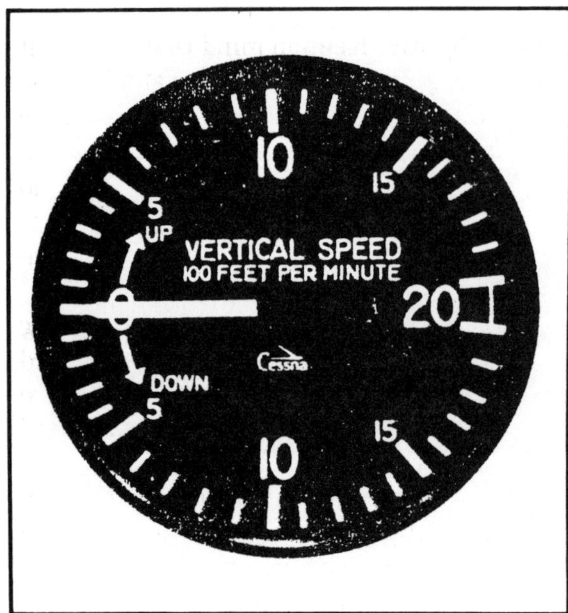

Fig. 2-15. Vertical speed indicator.

of the instrument is a diaphragm that reacts to changes in pressure occurring during climbs or descents. As the outside pressure changes, the instrument will show the rate of the change calibrated in hundreds of feet per minute.

The VSI is quick to show a vertical trend. It does, however, have a lag of six to nine seconds before indicating the correct *rate* of climb or descent. This lag occurs because of the delay necessary for pressure changes to stabilize inside the instrument.

START, TAXI, AND RUN-UP

So much for admiring the panel. Now let's crank it up . . . by the numbers. Checklists are used to determine if the aircraft systems are set properly and operating normally. Another of those all-important early habits to start now is to always use a checklist.

Before Start

- ☐ Brakes tested and hold—Pedals should feel firm; should not be or become soft and mushy.
- ☐ Seats adjusted and seatbelts fastened—All controls within reach without undue movement. Seatbelts and harnesses adjusted for snug fit (no slack). Make sure all belts are inside aircraft (not hanging out door).
- ☐ Radios and electrical OFF—Because of the large power requirement of the starter, other components could be damaged while starting.
- ☐ Fuel value ON—Placed to on, both, or fullest tank as appropriate for the make and model.
- ☐ Mixture RICH—Mixture knob full forward.
- ☐ Prime if required—Most aircraft engines will need priming to help start for the first time each day. The amount of prime will vary with make and model and with temperature. Your instructor will explain this to you. If the engine has been run in the previous several hours, priming will not be needed.
- ☐ Throttle 1/4 inch OPEN—With throttle off, place tip of index finger 1/4″ back from friction lock, then move throttle forward until finger just touches. Too much or too little throttle will prevent the engine from starting.
- ☐ Area "clear" (voice command)—Look around the aircraft to ensure that no one is in the vicinity of the prop; then, in a reasonably loud voice (not a scream), say, "Clear!"
- ☐ Master ON—Turn both halves of the master switch on. Between sunset and sunrise,this is the point where the position lights are turned on.

- ☐ Magnetos ON—Rotate key to the BOTH position.
- ☐ Engine START—Rotate key to start position. As engine starts, release key. If engine fails to start within 15 seconds, stop cranking to allow starter to cool and return to Step 4 and try again. If engine does not start on the second attempt, get assistance from an instructor. Don't continue trying to start until the battery is dead.

Post Start

- ☐ Engine at low rpm until oil pressure is indicating. Keep RPM at 600-800 until oil pressure is above the lower redline. If oil pressure does not come up in approximately 20 seconds after start (45 seconds if the temperature outside is below 30° F.), shut engine down and have it checked.
- ☐ Lean mixture as recommended.
- ☐ Align DG with the magnetic compass.
- ☐ Flaps UP—Retract flaps prior to taxi.
- ☐ Radios ON—Turn radios on, set volume, squelch, and frequencies.

Ground Operations

Shortly after starting to taxi the aircraft, test the brakes to determine that they are operating properly and that both pedals have approximately the same amount of travel. Now is the time to ensure that you have brakes, rather than waiting until they are needed and then finding that you don't have any.

The taxi speed for safe operation depends upon your surroundings. If you're in the ramp area where there are lots of people and other airplanes, you should never use a taxi speed above that of a leisurely walk. If you are out on an open taxiway, the speed may be increased slightly. Keep in mind that you want to have the aircraft under control at all times. Water or snow on the taxiway will also have a bearing on your taxi speed. Aircraft have less effective braking action than cars, and no power going directly to those wheels. Obstacles such as hangars can hide a potential hazard from your view.

Great car must be taken to avoid dragging the brakes while taxiing. Excessive heat and brake wear will result. Since you don't drive your car with your foot on the brake, don't taxi an airplane that way. Use the rudder and steerable nosewheel for directional control with momentary application of one brake to help with sharp corners. Brakes on an airplane are used to hold it from moving while doing the normal run-up and stopping while at slow speeds. The tires on your aircraft are quite small, and excessive braking can rally have an exaggerated effect on tire wear.

The Pre-Takeoff Checklist

- ☐ Set idle for smooth operation—Lowest rpm that still results in a smooth-running engine (1000 rpm is to high).
- ☐ Controls FREE—Movement and response.
- ☐ Check and set all instruments.
- ☐ Check fuel valve ON.
- ☐ Check flaps and flap indicator.
- ☐ Trim TAKEOFF position.
- ☐ Run-Up:
  - —Throttle setting 1700 rpm.
  - —Engine instruments within green arcs.
  - —Suction within green arc.
  - —Mixture SET for density altitude.
  - —Carb heat check (for tachometer response,then off).
  - —Magnetos check.
  - —Return magnetos to BOTH.

- ☐ Throttle friction lock set.
- ☐ Scan checklist for missed items.
- ☐ In case of delay, increase rpm to prevent plug fouling.
- ☐ Check doors LOCKED.

The run-up is done just before takeoff to assure that the aircraft is ready to fly. A little practice will make this routine quick and easy, but for now, let's do it a second time for detailed understanding.

Controls

All the controls are moved through their full range to check for any binding or possible loose play, and to see that the control surface is deflected in the proper direction. With the yoke rolled fully to the left, the ailerons are checked for the proper direction (left up and right down) while the elevator is raised to the full up position and checked. Then the ailerons are turned to the full right position and checked, while the elevator is lowered to the full down position.

With the aileron remaining in the full right position, the elevator is again raised to the full up position and checked. The aileron is then returned to the full left position and the elevator returned to the full down position. This ensures that the ailerons do not bind at full deflection left and right.

The rudder is checked separately for full deflection both left and right with no binding. If the rudder is hard to move in one direction, then the nosewheel probably isn't centered—center the nosewheel and try again.

Instrument Checks and Settings

Starting in the upper left-hand corner, all the instruments are checked for proper operation and/or readings. The sequence listed below may not be the exact sequence that you use, but all instruments must be covered.

- ☐ Airspeed indicator—Reading zero.
- ☐ Attitude indicator—Fully erect and miniature airplane set on the horizon.
- ☐ Clock—Operating and set.
- ☐ Altimeter—Set for current altimeter setting or field elevation. Note the pressure in the window.
- ☐ Vertical speed indicator—Reading zero. If not 0, this error will remain constant and will have to be taken into consideration when airborne.
- ☐ Heading indicator (DG)—Set to magnetic compass.
- ☐ Turn coordinator—Wings level and ball centered.
- ☐ Tachometer—Operating properly at 1000 rpm.
- ☐ Suction gauge—In the green (4.6—5.4 inches).
- ☐ Ammeter—Should show zero or possibly a slight charge. The landing lights are turned on for 1-2 seconds to determine if alternator will support additional load.
- ☐ Fuel gauges—Confirm preflight visual check.
- ☐ Oil pressure—Green arc.
- ☐ Oil temperature—Green arc. If it is the first flight of the day, it may not show a reading until after takeoff.
- ☐ Fuses—In place; spares available.
- ☐ Fuel—Selector valve set for takeoff.

Flaps

The flaps are checked for proper operation and set to the desired position for takeoff. To prepare you for the time when you start using flaps, they should be stopped at each 10 degrees on the way down. On the way up, they should be retracted from full directly to 20 degrees and stopped, up to 10 degrees, then full up or retracted. This routine is so that the flaps are being extended exactly as you will be doing while in an approach to landing and are retracted as though making a full-flap go-around or aborted landing. After doing this a few times, you will get a sense of timing for extending and retracting the flaps, which will make you less dependent on the flap indicator. Then you can spend more time paying attention to the aircraft and flying it properly while still knowing how much flap is extended.

Trim

The trim control is moved both up and down and the trim tab is checked visually to see that it moves accordingly. The trim tab is then adjusted for the takeoff and the trim indicator is checked.

Run-Up

Holding the brakes, the throttle is slowly advanced to 1700 rpm, where we will check that the engine instruments are within their green arcs. Then the mixture is *slowly* pulled out to lean the engine to its peak rpm. Since full power will be used for takeoff, we'll next richen the mixture halfway to provide cooling of the engine to prevent detonation while at full throttle.

Carb heat is pulled for approximately five seconds to check for operation and icing. Normally, rpm will drop a bit and remain stable until the heat is turned off. If there is ice the engine may run rough, but the primary indication will be that after the initial drop in rpm, it will slowly climb part way back up as the ice melts. Then, after returning the carb heat to the closed of OFF position, the rpm will be higher than before the heat was applied. If carburetor ice was detected, we will reset the mixture before proceeding. Extended use of the carburetor heat on the ground is not recommended due to unfiltered air being drawn into the engine.

The magnetos are now checked separately to ensure that they are both functioning. Move the ignition switch first to one mag and note the rpm on the tachometer. Then move the switch back to BOTH to clear the other set of spark plugs, then over to the other mag. Note the rpm, and return to BOTH. The difference between the two mags operated individually should not be more than 75 rpm. An excessive difference *or* the absence of *any* drop on one mag should be checked out before flight. It should be noted that the individual drop for each magneto should not exceed 125 rpm!

Friction Lock

The throttle friction lock should be set tight enough that it won't vibrate out, but loose enough that it doesn't take a great effort to move the throttle.

Ready For Takeoff

Now the checklist is completed and the preliminaries are nearly over. At this time your instructor will coach you through the takeoff.

THE TAKEOFF

This "normal" takeoff is probably the technique that you will use most, unless you specialize in bush-type flying out of short, unpaved strips.

After the aircraft is aligned with the centerline of the runway, power is smoothly applied until full rpm is being developed. The time lapse from when you start to add power until max power is reached should be from three to five seconds. As high power is developed at a slow forward speed, care must be taken to keep the aircraft headed down the centerline with the rudder pedals. In a no-wind condition you'll tend to turn slightly to the left because of the effects of torque.

As soon as practical, an rpm check must be made to see if the engine is developing takeoff power. If there is some doubt in your mind as to whether the engine is developing full power, the takeoff can be aborted by smoothly closing the throttle early enough to still be able to stop the aircraft on the runway.

It is important that the elevator be allowed to slipstream or even be held slightly up to reduce the load on the nosewheel. As the aircraft continues to accelerate, the control surfaces become more and more effective. You will soon be able to judge by pressure on the elevator when there is sufficient control to pitch the airplane gently upward.

Near 55 KIAS (V_x), the nosewheel should come off the ground; this is the takeoff attitude. With enough speed to fly, the airplane will fly all by itself. Our next job is to accelerate to the best rate of climb speed (V_y) of 65 to 67 KIAS, and maintain this airspeed until a safe altitude is reached.

You will soon discover that the aircraft accelerates better near the ground because of what is called *ground effect*, so we will lower the nose just slightly to prevent the aircraft from climbing out of ground effect until we reach V_y. This will be the normal climb speed for about the first 400 feet of altitude at this airport. (Keep in mind that at other airports you may be flying right over a city immediately after takeoff, so this speed would be held longer.)

As soon as a safe altitude is reached we will again lower the nose slightly and accelerate to "cruise climb," where we may reduce the power a bit. The climb is then continued at this speed until the desired cruising altitude is reached.

If you wish to depart the pattern, a turn to crosswind should be made. After clearing your flight path, turn 45° in the opposite direction of the traffic pattern in order to depart that pattern.

Climb Speed Data And Definitions

Our first important airspeed is known as *best angle of climb* airspeed (V_x). This means that the aircraft will climb to the greatest altitude for any given horizontal distance traveled. This is the speed that is used with short-field takeoffs over an obstacle.

Second, and the one referred to in the above narrative about climbout, is the *best rate of climb* airspeed (V_y). This is the speed at which the aircraft will climb to the greatest altitude for any given time period. It's an airspeed that will allow you to get away from the ground in the shortest possible time.

Another climb speed is the *cruise climb* (no V designation for this one), which in the C152 is 5-7 kts above V_y. This is considered a normal climb speed after all obstacles have been cleared and a safe altitude has been reached. This speed will allow for better engine cooling, greater visibility, and a lower noise level than the other two climb speeds and is used primarily for these reasons.

Figure 2-16 represents the three climb speeds and what they do for you. The vertical scale is altitude and the horizontal scale is ground or horizontal distance. The liftoff points in all three cases are assumed to be the same and the length of each flight path is the same amount of time.

Notice that at V_y the most altitude is attained per unit of time and therefore the VSI would give the highest reading possible under the given conditions. At V_x both the amount of altitude gained over time and the distance traveled over the ground are less than at V_y; however, the height above the terrain is

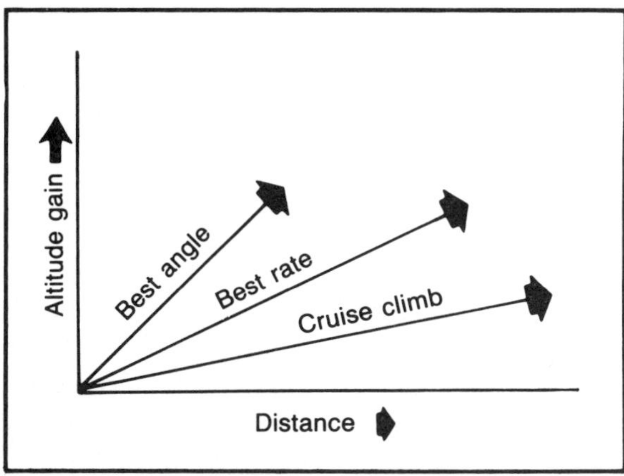

Fig. 2-16. The climb speeds.

greatest when the aircraft reaches any point. Even though the rate of climb is less at V_x, the aircraft has a longer period of time to climb because of the slower airspeed and therefore will have more altitude over a given obstacle. At cruise climb, the aircraft will again have less altitude but will travel farther over the ground for the same time period.

Your First Look at Aerodynamics

Any "airfoil shape" (wing, propeller blade,etc.) produces lift by a reduction of pressure over its curved surface. This is accomplished at least partially by what we call the *angle of attack*. This is the angle measured between the relative wind and the chord line (Fig. 2-17). *Relative wind* is the airflow crated by the aircraft's forward movement, parallel and opposite the direction of flight. The *chord* is an imaginary straight line from leading edge to trailing edge of the airfoil.

The pilot's primary control of the angle of attack is through the elevators (pitch attitude). To increase lift (up to the point of stall), the angle of attack is increased as we just did in the climb speeds discussion. V_x is a higher pitch attitude and angle of attack, which allows us to fly (and climb) at a slightly slower airspeed than V_y. Cruise climb is a slightly lower angle of attack, allowing more speed for any given power setting; this extra speed will regain whatever lift we gave up by using the lower angle of attack.

Climbout and Left Turning Tendencies

We've finally passed the end of the runway and are on our way. Something's wrong—the airplane wants to wander off to the left. Why? Let's take a look:

In most American aircraft, the propeller rotates clockwise as viewed from inside the cockpit. In level flight at cruise airspeed, the angle of attack on both blades is equal and the propeller lifts or pulls equally on each side (Fig. 2-18A). However, when the aircraft is in a nose-high attitude with power on, as in slow flight or a climb, the angle of attack on the ascending blade (pilot's left) is decreased and the angle of attack on the descending blade (pilot's right) is increased (Fig. 2-18B). Since

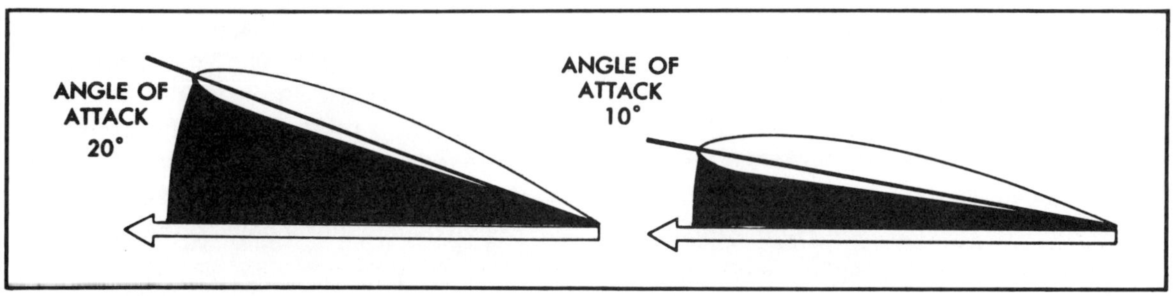

Fig. 2-17. Angle of attack.

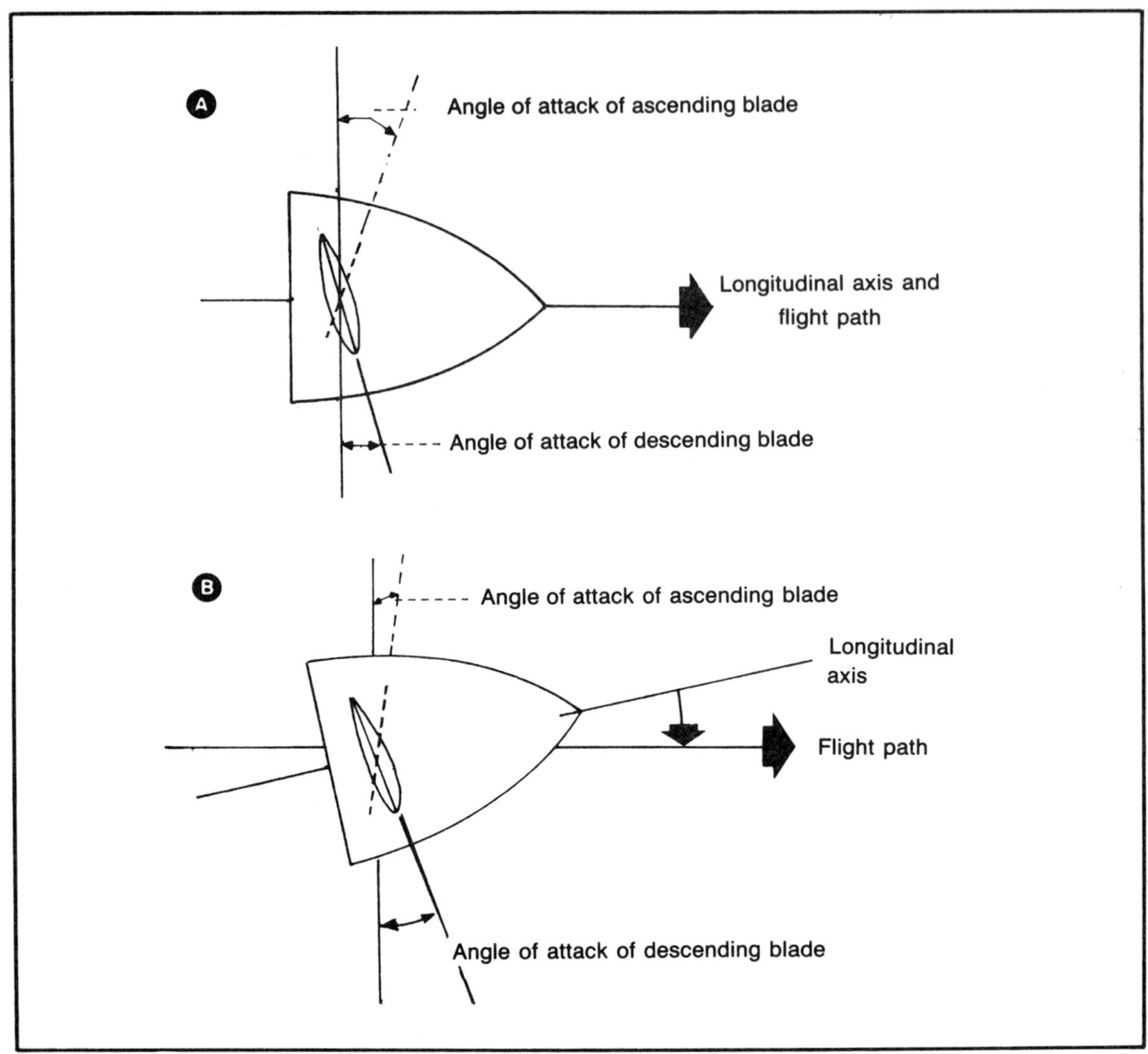

Fig. 2-18. P-factor.

the speed of the relative wind is equal on both blades, the right blade will be producing more lift than the left blade. This effect is called *P-factor* and will tend to yaw the aircraft to the left. Your correction is by right rudder pressure. *Any time the angle of attack is increased and power is on, add right rudder.*

Also, ever since Sir Isaac Newton declared his Third Law to be in effect, those clockwise spinning propellers have been tending to roll aircraft into a counterclockwise (left) bank. This is called *torque,* and it is present whenever the engine is producing power. For cruise flight, the manufacturer thoughtfully designed into the airplane a correction so that the pilot won't have to constantly hold control pressure. This can be done in a number of ways, such as increasing the *angle of incidence* (between the chord line and the longitudinal axis) on the left wing or by offsetting the engine two or three degrees to the right. Keep in mind that these corrections are made to stabilize the aircraft at the speed it was designed to be used at most—cruise. In a climb, the correction is less effective because of the lower airspeed, so we must use a slight right aileron pressure throughout the climb, coordinated with the right rudder pressure covering the P-factor.

Of course, your flight instructor is coaching you through all of this and, on the first few flights, he will probably direct you out of the traffic pattern and away from the airport to a practice area. As you leave the pattern, he may also have you make your first call on the radio over a frequency assigned to something we'll call "Weld County Unicom."

Unicom is a non-official but very handy service provided by most public airports. Ours is typical in that someone at the airport terminal monitors the frequency of 122.8 MegaHertz (MHz) and advises incoming pilots of the winds and which runway is in use by other aircraft in the landing pattern. Unicom can also be used for other services the pilot might need, such as, "Please call my office at this number . . ." "What's the price of fuel?" or "Order me a pizza." Remember that Unicom is not the FAA, or what we call Air Traffic Control (ATC).

The inbound traffic calls Unicom on 122.8 MHz, but we're departing, so we'll just use their frequency to advise other traffic of our intentions, such as: "Weld County Traffic, Cessna 12345 departing the pattern to the east"—or something like that, as appropriate.

In the Traffic Pattern

Runways are numbered in relation to their magnetic heading. Starting at north, we have 0 or 360 degrees. Clockwise, the numbers increase as in 090 for east, 180 for south, and 270 for west. Drop the last zero and you have the runway number, such as 3, 6, 9, 12, 27, etc. (Fig. 2-19). If your actual runway heading

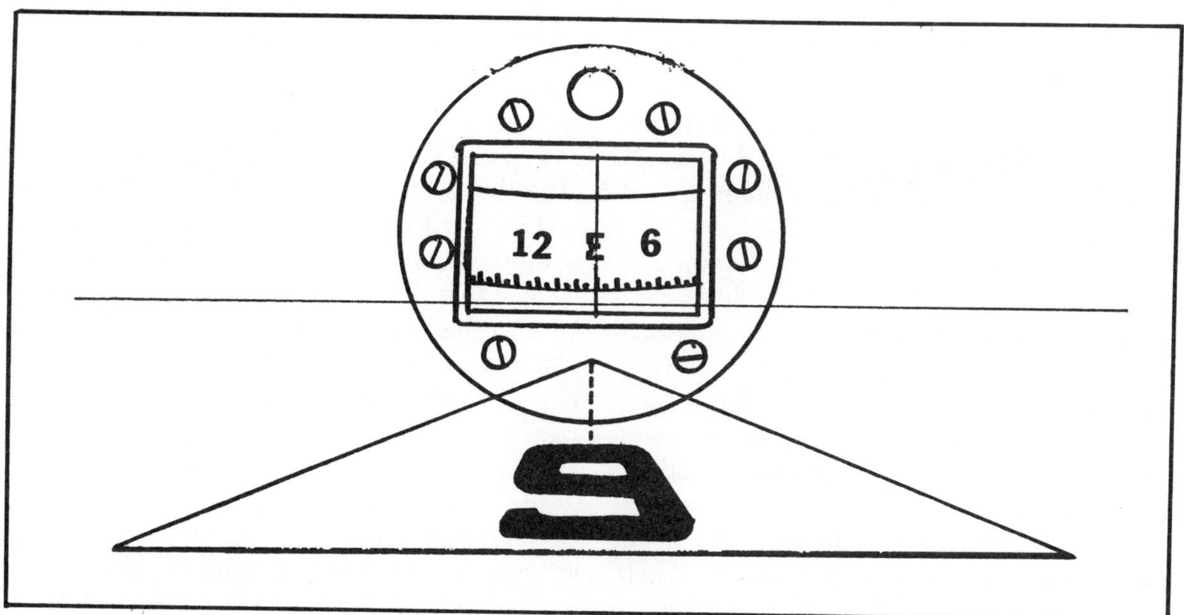

Fig. 2-19. Compass heading and runway number.

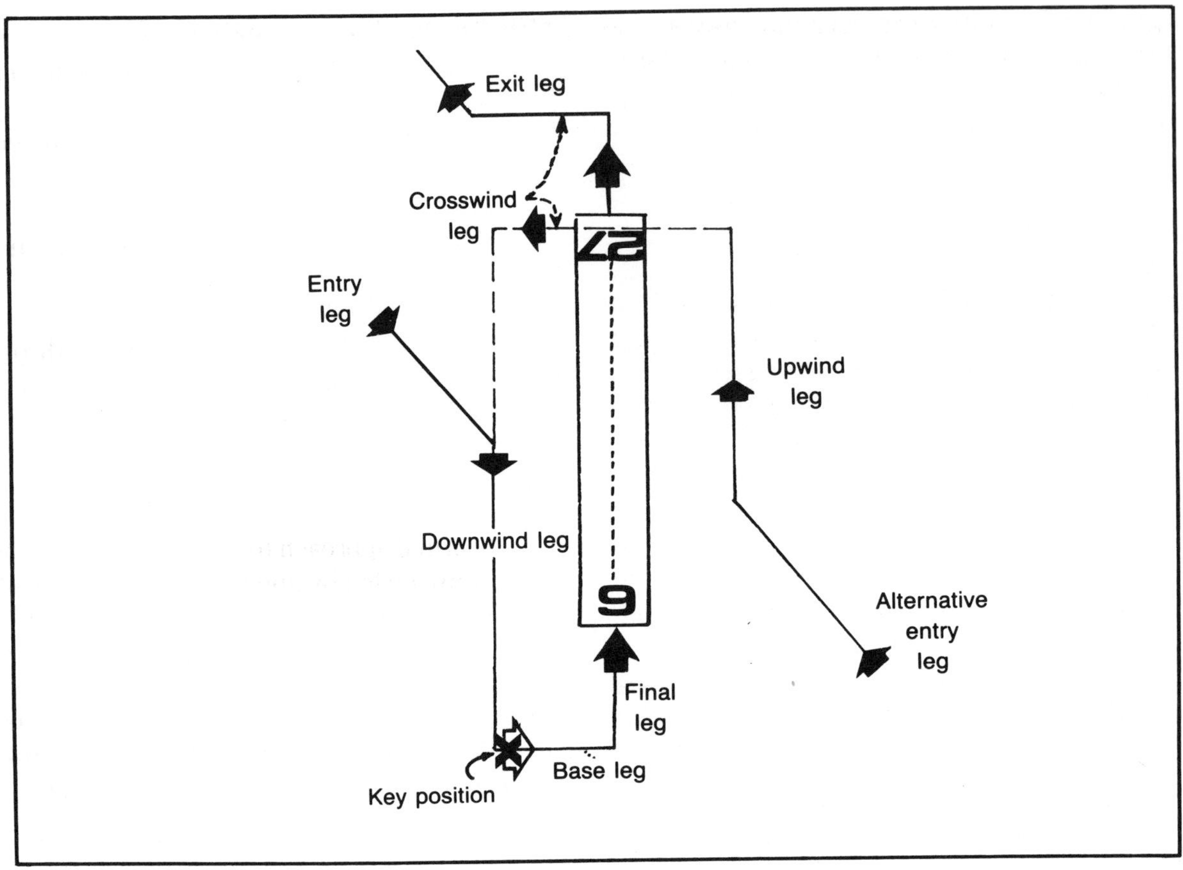

Fig. 2-20. Left (standard) pattern to runway 09.

is 034 degrees, it will be rounded off to the nearest whole ten degrees and will become 03. If the heading were actually 035 degrees, the runway number would be 04.

The traffic at an airport would be disorganized and unpredictable if there were no way for a pilot to know how to approach the field to load. The rectangular pattern (Fig. 2-20) has evolved through tradition and is accepted as a normal procedure while preparing to land. Notice that all the turns are made to the left, so this is called a *left-hand pattern.* Left patterns (or "left traffic") are used whenever possible because the runway is more visible to the pilot sitting in the left seat. It is standard at all airports unless designated otherwise. (However, at a controlled airport, the control tower may issue instructions for the pilot to alter this pattern in the interest of expediting traffic flow.) You should memorize the names of all the legs of the traffic pattern so you will be able to visualize the pattern once you know which runway is in use. This will also help you place other aircraft when you hear them announcing their positions in the pattern over the frequency in use.

Pattern entries will be made at 45 degrees to the downwind leg or, if more convenient, 45 degrees to the upwind leg. From the entry, we will fly the remainder of the pattern to touch down as shown in Figs. 2-20 and 2-21. If you're approaching a strange airport and aren't sure of its exact location, stay at least 1500 feet above ground level (AGL), Normal traffic pattern altitude is 800 feet AGL, so this will keep you clear of traffic until you can plan your entry into the pattern. From this altitude, you can determine if there is a right-hand pattern in use yet still be able to read the wind indicators.

But for now let's say that you've not left the airport at all and are remaining in the pattern for a series of practice landings or "touch and goes." After each takeoff, maintain a ground track that will be an extension of the runway centerline. After reaching a safe altitude, clear yourself for traffic, then turn left to change the ground track by 90 degrees.

Staying in the pattern involves another 90 degree left turn onto the downwind leg. Be sure to look first before making *any* turn. It's not only professional, it will probably someday save your life!

While flying downwind, we will run through another of those checklists:

The Pre-landing Checklist

- ☐ G—Gas check (tanks and selector valve).
- ☐ U—Undercarriage (landing gear) down and locked.
- ☐ M—Mixture set for landing.
- ☐ P—Propeller set for landing (if applicable).
- ☐ S—Seat belts secure.
- ☐ Carburetor heat on just before throttle reduction.
- ☐ Flaps as needed below V_{fe}.

LANDING

A normal approach to landing is started on the downwind leg by applying carburetor heat about halfway down the runway to prevent any ice formation. Opposite the point of intended landing (the white line, or threshold), the throttle is smoothly reduced to approach power (usually this will be about 1500 to 1700 rpm).

Now we lower the pitch very slightly.

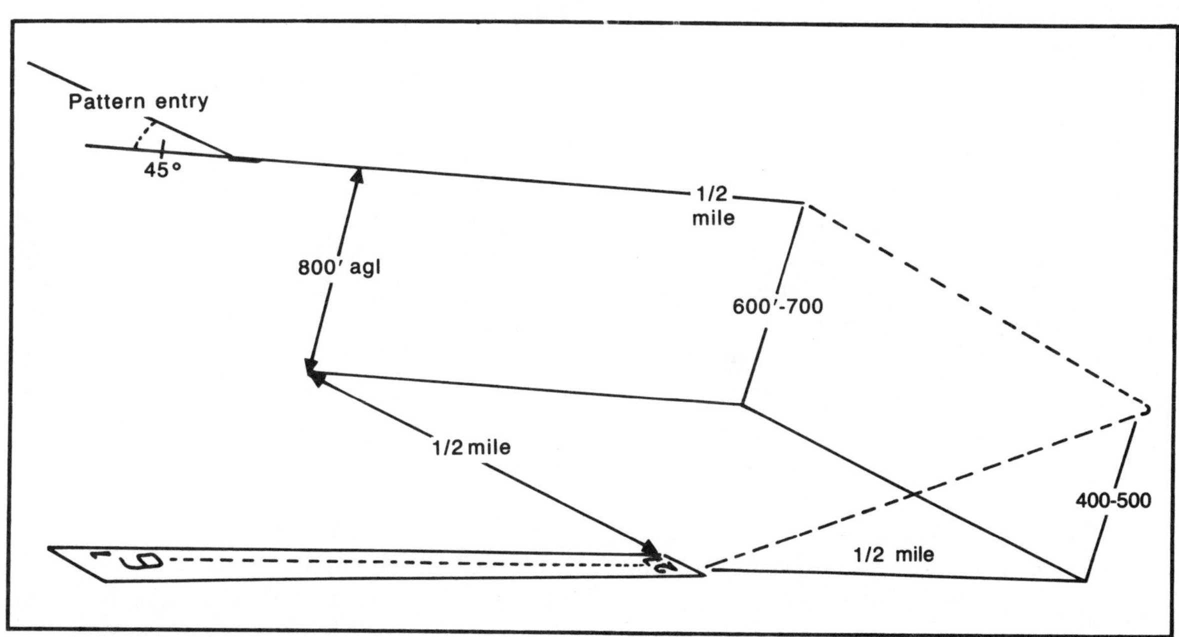

Fig. 2-21. Right-hand pattern.

Because of the rather large power reduction, it is quite likely that we will have to hold back pressure on the yoke to control the pitch attitude. We will then use the trim to remove the pressure we are holding and stabilize the aircraft at 65-70 knots.

When the end of the runway appears to be about 45 degrees behind you, it is time for the turn to base leg. The wing must be lifted prior to the turn so you can see if any traffic is turning inside you or overtaking you from that side. A good rule is that 30 degrees of bank is the maximum to use while in the pattern and 20 to 25 degrees will produce a comfortable rate of turn. If the turn is too shallow, you are blind for along period of time and the radius of turn will be too great, which will cause you to overshot the centerline of the runway.

As you roll out of the bank for the turn to base, the lower wing should be raised above level to quickly check for traffic turning inside you. Once this is accomplished, we are at the *key position,* where we can determine if any large changes in power must be used to adjust our angle of descent. After clearing the area to the right for any traffic on final for a straight-in approach, we can begin our turn to final leg.

Don't be afraid to lean forward in the seat and watch the runway through the windshield so you can line up on the extended centerline of the runway. Here, slight variations of the bank will increase or decrease the radius of turn so we can follow our desired ground track. Still keeping the airspeed constant, we can watch the runway; if it appears to be coming under us, we are overshooting and must reduce power to increase our angle of descent. If the touchdown area does not appear to move but only gets larger, we are on the correct glide angle, but if it appears to move up, we are too low and must add power. Remember, if the airspeed remains constant, power controls the altitude.

When we see that we have the runway made, we can close the throttle completely and a slow, smooth flare is carried out to stop the rate of descent. As the aircraft slowly decelerates, we must continue to increase the back pressure to maintain the pitch attitude and keep from settling onto the runway at an excessive ground speed. During this transition period your eyes should move out in front of the aircraft to the same distance and focus you would use if you were driving your car down a highway at 55-60 mph. If you look too close to the aircraft, there will be a great tendency to flare too close to the aircraft, there will be a great tendency to flare too high because of the extreme sensation of speed, whereas if you look too far away, you will not be able to detect any change in your altitude or height above the runway, and will tend to flare too late. We can keep raising the nose as long as the aircraft does not climb back up until we reach the landing attitude.

If we maintain this attitude by slowly continuing the back pressure as the airspeed bleeds off, the aircraft will slowly settle to the runway as it continues to decelerate.

After the main wheels are on the ground, maintain the same pitch attitude with the elevators to prevent the nosewheel from coming down until the aircraft gets so slow that you can no longer maintain this pitch attitude, and the nose too will slowly settle to the runway. It is important to realize that we really don't land the aircraft; we only reach a given attitude and power combination at a specific height above the runway and the aircraft lands itself. While we are approaching this attitude and height, we must also keep the aircraft flying down the centerline of the runway and keep the longitudinal axis or heading parallel with the runway to prevent any sideloads on the landing gear.

SHUTDOWN

- ☐ Radios and all electrical systems OFF.
- ☐ Throttle 1000 rpm.
- ☐ Mixture IDLE CUTOFF.
- ☐ Throttle OFF when engine stops firing. (This can be detected by a change in the sound of the engine.)
- ☐ Magnetos OFF after engine completely stops.
- ☐ Master OFF.
- ☐ Control Lock IN.
- ☐ Aircraft tied down; chock wheels.

(*Note:* Shutdown at night is the same as day except the navigation lights stay on until the magnetos are off.)

PROBLEMS AND LEGALITIES

Emergencies caused by aircraft or engine malfunctions are extremely rare, especially if proper maintenance and preflight inspections are practiced. Should an emergency arise, your best tools will be a knowledge of the aircraft and how its systems operate, an ability to keep calm and cope with the situation, and application of the basic guidelines in this section as necessary to correct the problem.

First, it would be a good idea for you to study the engine failure checklist for complete familiarity. If it should ever happen to you, there may not be sufficient time to get out the checklist—and your mental state won't be helping your memory. Practice dry runs on the ground until you can do it in a daze; someday you may have to do just that!

Engine Failure Checklist

- ☐ Carb heat ON.
- ☐ Close throttle.
- ☐ Mixture RICH.
- ☐ Gas ON.
- ☐ Primer In or LOCKED.
- ☐ Mags—check on BOTH.
- ☐ Proceed to landing site.
- ☐ Best glide speed.
- ☐ Open throttle to try for restart.
- ☐ Radio call (Emergency is 121.5 MHz).
- ☐ Set up an approach to the field so that the touchdown point is 1/3 to 1/2 into the field.
- ☐ All unnecessary equipment OFF.
- ☐ Unlatch door before touchdown.
- ☐ Night Emergency: Full nose-up trim; full flaps.

Electrical System Malfunctions

Electrical malfunctions will usually be experienced as insufficient rate of charge or excessive rate of charge.

If the ammeter indicates a continuous discharge rate in flight, the alternator is not supplying power to the system and should be shut down since the alternator field circuit may be placing an unnecessary load on the battery. All non-essential equipment should be turned off, the alternator side of the split master switch should be turned off, and the flight terminated as soon as practical.

After periods of prolonged engine starting or heavy electrical usage at low engine speeds, the battery condition will be low enough to accept above-normal charging during the initial part of a flight. However, after 30 minutes of cruising flight, the ammeter should be indicating less than two needle widths of charging current. If the charging rate remains *above* this value on a long flight, it is possible that the battery or other components of the electrical system could be damaged. To prevent this, the alternator side of the split master switch should be turned off. The flight should be terminated and/or the current drain on the battery minimized as soon as practical because the battery can supply the electrical system for only a limited time period. If it becomes apparent that the battery is getting too low to operate the electrical system, the alternator switch can be turned back on for several minutes at a time until the battery is partially recharged. If the emergency occurs at night,the alternator switch can be turned back on just before landing lights and flaps are needed for landing.

Loss of Oil Pressure

If low oil pressure is accompanied by normal oil temperature, there is a possibility that the oil pressure gauge is malfunctioning. In any case, a landing at the nearest airport would be advisable to inspect the source of the trouble.

If a total loss of oil pressure is accompanied by a sudden rise in oil temperature, there is good reason to suspect an engine failure is imminent. Reduce engine power immediately and select a suitable forced landing field. Leave the engine running at low power during the approach, using only the minimum power required to reach the desired touchdown spot.

Magneto Failure

A sudden engine roughness or misfiring is usually evidence of magneto problems. Switching from BOTH to either LEFT or RIGHT ignition switch position will identify which magneto is malfunctioning. Select different power settings and richen the mixture to determine if continued operation on both magnetos is practicable. If not, switch to the good magneto and proceed to the nearest airport for repairs.

Other Aircraft

Generally, after you solo, you will be somewhat like you were when you first learned to drive a car: You heard all the gory stories about accidents, and watched all those safety films with titles like *Speed Kills*, and *Drag Racing Daredevils!*—all of which were necessary to scare Hell out of you and make you a sane driver. Flight training has its share of "scarem" moments, carefully built into your program by an instructor, to keep you constantly alert and forever in the "good flying habits" mode. It has been said that flying is "nine hundred and ninety-nine hours of sheer boredom laced with one hour of absolute terror!" this is probably stretching the truth a bit, but it does prove a point. Flying has its moments and one of them is the "near miss"—an "almost" accident that could have been a midair collision. Let's take a look at the *Airman's Information Manual:*

"A high percentage of near midair collisions occur below 8,000 feet AGL and within 30 miles of an airport. When operating VFR (by Visual Flight Rules) in these highly congested areas, whether you intend to land at an airport within the area or are just flying through, it is recommended that extra vigilance be maintained and that you monitor an appropriate control frequency . . ." (Paragraph 571, *Airman's Information Manual,* FAA)

"If you think another aircraft is to close to you, give way instead of waiting for the other pilot to respect the right-of-way to which you may be entitled. It is a lot safer to pursue the right-of-way angle after you have completed your flight." (Paragraph 570 (d), *Airman's Information Manual,* FAA)

What the FAA is saying is that it's much better to take immediate evasive action *regardless* of who has the right-of-way. Later, when you get back to the airport, you can analyze the legalities to your heart's content. With that understood, let's take a look at the Federal Aviation Regulations on right-of-way.

Right-of-Way (FAR 91.67)

(A) GENERAL: When weather conditions permit, regardless of whether an operation is conducted under Instrument Flight Rules or visual Flight Rules, vigilance shall be maintained by each person operating an aircraft so as to see and avoid other aircraft in compliance with this section. When a rule of this section gives another aircraft the right-of-way, he shall give way to that aircraft and may not pass over, under, or ahead of it, unless well clear.

(B) IN DISTRESS: An aircraft in distress has the right-of-way over all other air traffic.

(C) CONVERGING: When aircraft of the same category are converging at approximately the same altitude (except head-on, or nearly so), the aircraft to the other's right has the right-of-way (Fig. 2-22).

If the aircraft are of different categories:

- □ A balloon has the right-of-way over any other category of aircraft;
- □ A glider has the right-of-way over an airship, airplane, or rotorcraft;

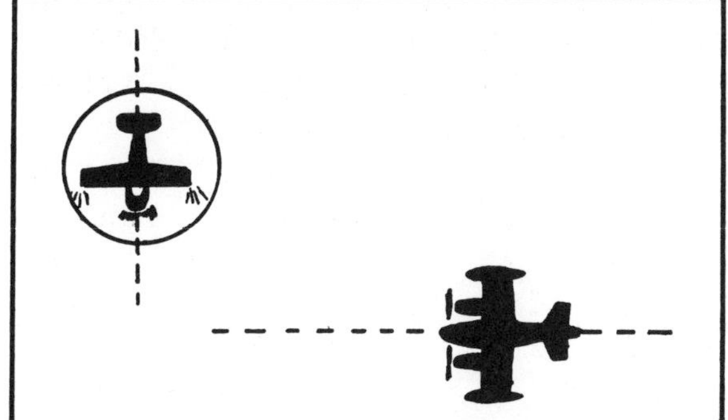

Fig. 2-22. Circled airplane has right-of-way.

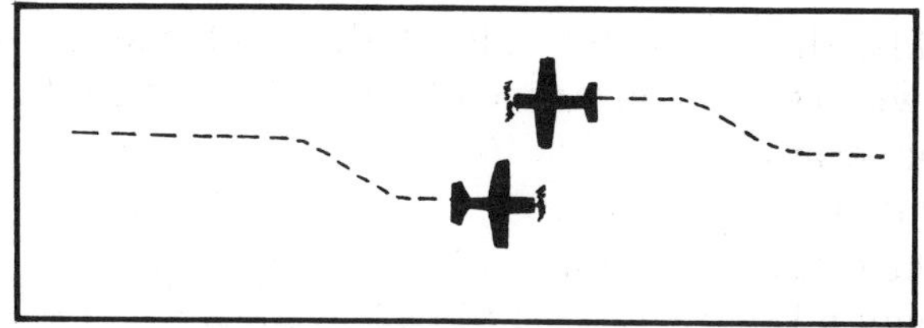

Fig. 2-23. Right-of-way, head-on.

- □ An airship has the right-of-way over an airplane or rotorcraft.

However, an aircraft towing or refueling another aircraft has the right-of-way over all engine - driven aircraft.

(D) APPROACHING HEAD-ON: When aircraft are approaching each other head-on, or nearly so, each pilot of each aircraft shall alter course to the right to pass well clear (Fig. 2-23).

(E) OVERTAKING: Each aircraft that is being overtaken has the right-of-way and each pilot of an overtaking aircraft shall alter course to the right and pass well clear (Fig. 2-24).

(F) LANDING: Aircraft, while on final approach to land, or while landing, have the right-of-way over other aircraft in flight or operating on the surface (Fig. 2-15). When two or more aircraft are approaching an airport for the purpose of landing, the aircraft at the lower altitude has the right-of-way, but it shall not

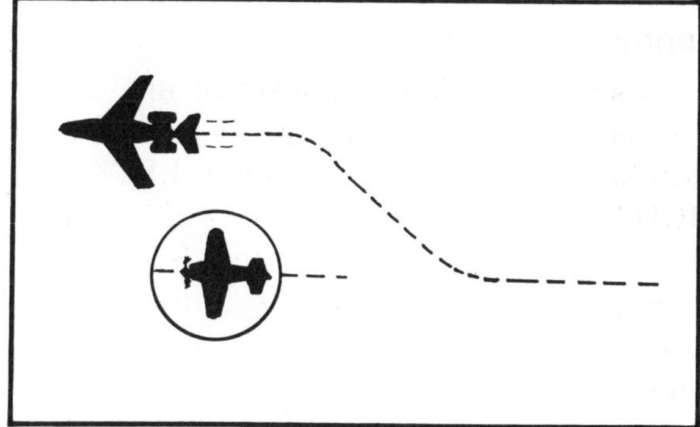

Fig. 2-24. Right-of-way, overtaking.

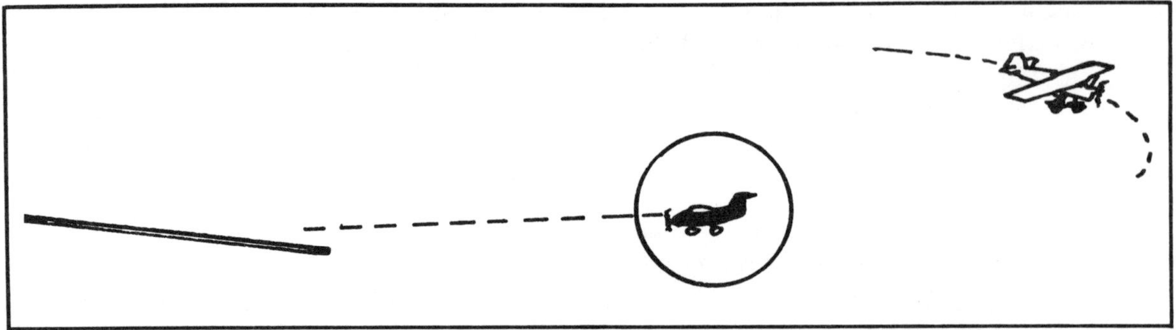

Fig. 2-25. Right-of-way, landing.

take advantage of this rule to cut in front of another which is on final approach to land, or to overtake that aircraft.

Another possible problem with other aircraft sharing your sky is the aerodynamic hazard of *wake turbulence*. Earlier we mentioned the low pressure over the top of the wing that is always present when lift is being produced. The higher-pressure air beneath the wing spills around the wingtip in an attempt to fill in that low to equalize the pressures.

This airflow around the tip, combined with the relative wind, creates *vortices* (or vortexes), which trail behind and below the aircraft (Fig. 2-26), creating severe turbulence.

You should always remember to avoid flying too close behind another airplane, especially when he is "low, slow, and heavy." This is because the high angle of attack he is using, along with his great weight, acts to increase the pressure difference above and below his wing. Be especially cautious about landing or taking off behind large aircraft. Since the vortices trail behind and below, you can easily visualize the general location of the invisible turbulence (Fig. 2-27) and plan your takeoff actions accordingly. Perhaps it would be even more advisable to wait three or four minutes for the vortices to dissipate.

Plan your approach to land long, beyond the "big guy's" touchdown point. His vortices will disappear once he's on the ground and the

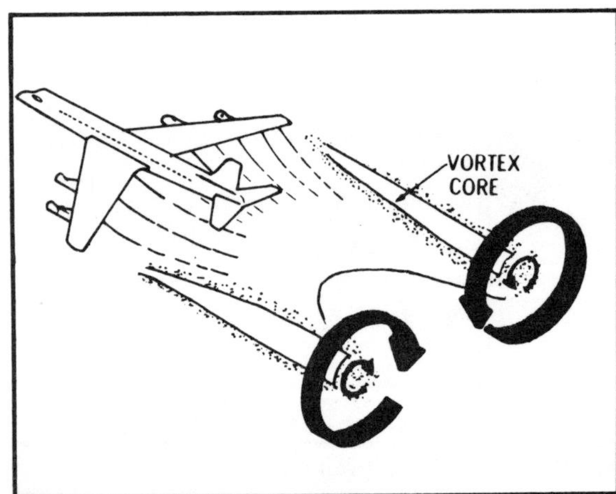

Fig. 2-26. Wingtip vortices.

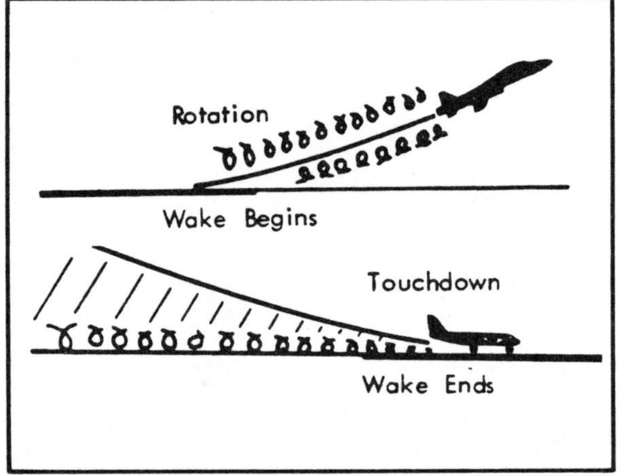

Fig. 2-27. Wake turbulence profile.

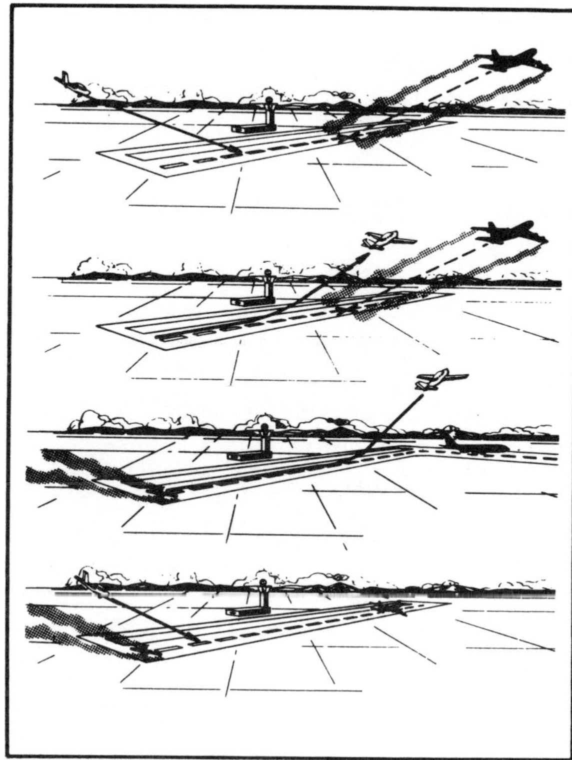

Fig. 2-28. Avoiding wingtip vortices.

weight transfers from his wings to his wheels. On takeoff, rotate before reaching his rotation point; then, since you probably can't outclimb him, turn slightly into any crosswind and make your climbout paralleling and upwind of the danger area (Fig. 2-28).

MORE BY THE BOOK

Here are some more important topics covered by the FARS.

FAR 91.3 Responsibility and Authority of the Pilot in Command

(a) The pilot in command of an aircraft is directly responsible for, and is the final authority as to, the operation of that aircraft.

(b) In an emergency requiring immediate action, the pilot in command may deviate from any (inflight regulation) to the extent required to meet that emergency.

(c) Each pilot in command who deviates from a rule under paragraph (b) of this section shall, upon the request of the Administrator, send a written report of that deviation to the Administrator.

FAR 91.7 Flight Crewmembers at Stations

(a) During takeoff and landing, and while en route, each required flight crewmember shall—

(1) Be at his station unless his absence is necessary in the performance of his duties in connection with the operation of the aircraft or in connection with his physiological needs; and

(2) Keep his seat belt fastened while at his station.

(b) . . . each flight crewmember . . . shall, during takeoff and landing, keep the shoulder harness fastened while at his station. This paragraph does not apply if—

(1) The seat at the crewmember's station is not equipped with a shoulder harness; or

(2) The crewmember would be unable to perform his required duties with the shoulder harness fastened.

FAR 91.11 Liquors and Drugs

(a) No person may act as a crewmember of a civil aircraft—

(1) Within eight hours after the consumption of any alcoholic beverage;

(2) While under the influence of alcohol; or

(3) While using any drug that affects his faculties in any way contrary to safety.

(b) Except in an emergency,no pilot of a civil aircraft may allow a person who is obviously under the influence of intoxicating liquors or drugs (except a medical patient under proper care) to be carried in that aircraft.

FAR 91.29 Civil Aircraft Airworthiness

(a) No person may operate a civil aircraft unless it is in an airworthy condition.

(b) The pilot in command of a civil aircraft is responsible for determining whether that aircraft is in condition for safe flight. He shall discontinue the flight when unairworthy mechanical or structural conditions occur.

FAR 91.79 Minimum Safe Altitudes; General

(a) ANYWHERE. An altitude allowing, if a power unit persons or property on the surface.

(b) OVER CONGESTED AREAS. Over any congested area of a city, town, or settlement, or over any open air assembly of persons, an altitude of 1,000 feet above the highest obstacle within a horizontal radius of 2,000 feet of the aircraft.

(c) OVER OTHER THAN CONGESTED AREAS. An altitude of 500 feet above the surface, except over open water or sparsely populated areas. In that case, the aircraft may not be operated closer than 500 feet to any person, vessel, vehicle, or structure.

FAR 91.89 Operation at Airports Without Control Towers

Each person operating an aircraft to or from an airport without an operating control tower shall—

(a) In the case of an airplane approaching to land, make all turns of that airplane to the left unless the airport displays approved light signals or visual markings indicating that turns should be made to the right, in which case the pilot shall make all turns to the right . . .

FAR 61.3 Requirements for Certificates,Ratings,and Authorizations.

(a) PILOT CERTIFICATE. No person may act as pilot in command or in any other capacity as a required pilot flight crewmember of a civil aircraft . . . unless he has in his personal possession a current pilot certificate issued to him under this Part.

(c) MEDICAL CERTIFICATE . . . no person may act as pilot in command or in any other capacity as a required pilot flight crewmember of an aircraft under a certificate issued to him under this Part, unless he has in his personal possession an appropriate current medical certificate . . .

FAR 61.87 Requirements for Solo Flight

(d) FLIGHT INSTRUCTOR ENDORSEMENTS. A student pilot may not operate an aircraft in solo flight unless his student pilot certificate is endorsed, and unless within the preceding 90 days his pilot logbook has been endorsed, by an authorized flight instructor who—

(1) Has given him instruction in the make and model of aircraft in which the solo flight is made;

(2) Finds that he has met the requirements of this section; and

(3) Finds that he is competent to make a safe solo flight in that aircraft.

FAR 61.89 General Limitations

(a) A student pilot may not act as pilot in command of an aircraft—

(1) That is carrying a passenger;

(2) That is carrying property for compensation or hire;

(3) For compensation or hire;

(4) In furtherance of a business . . .

FAR 61.93 Cross-Country Flight Requirements

(a) GENERAL. A student pilot may not operate an aircraft in a solo cross-country flight, nor may he, except in emergency, make a solo flight landing at any point other than the airport of takeoff, until he meets the requirements of this section . . . As used in this section, the term cross-country flight means a flight beyond a radius of 25 nautical miles from the point of takeoff.

Chapter 3

Navigation

When you drive out to the local airport on a "clear as a bell" Sunday, you will see a bevy of "touch-and-goers" practicing landings for the inevitable checkride. The unusual thing about it is that some pilots never seem to "graduate" from this stage of the game. They spend a large portion of their flying careers covering the same real estate weekend after weekend.

The Sunday flier is not to be criticized; when you think of the cost of flying today, you will have to admit that about all some of us can afford is local flying—$55 per hour for a Skyhawk is a pretty steep price to pay just to "go fly for a while." If we want to venture away from the local scenery, the price goes up considerably.

However, there comes a time when we want to spend a few dollars just to go visit relatives or take a business vacation. Our rusty skills are put on the line and we know that it's going to take more than just a glance through some old ground school book to get current in cross-country techniques. The point is, *learn it right the first time;* then, if you don't get to fly "XC" very often, discuss your needs with a flight instructor and even get some flying time with a CFI to rush up on those rusty skills. You might even take an experienced, current instructor along with you on a cross-country just to polish the rough edges. The cost? Hey, your life's at stake; what's that worth?

THE ELECTRONIC FLIGHT COMPUTER

The old "whiz-wheel" type flight computer was a wonderful device for solving navigation problems, but the handwriting is on the wall and the electronic calculator, preprogrammed for flight navigation problems, is the "now" method. We will be using the Jeppesen AVSTAR, which is rapidly becoming the most widely used model around (Fig. 3-1). (Thanks

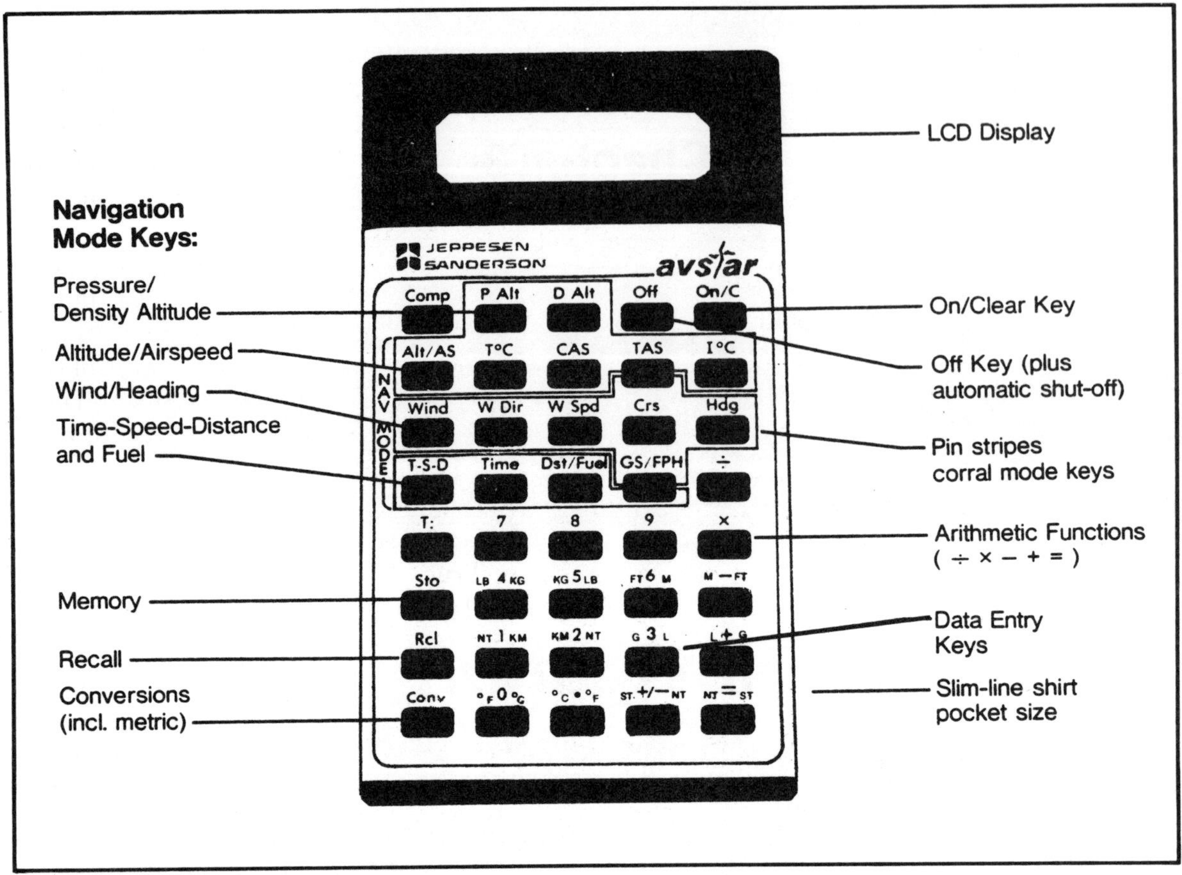

Fig. 3-1. Jeppeson Sandersen AVSTAR.

to Jeppesen Sanderson, 55 Inverness Drive, Englewood, Colorado, for permission to use excerpts from their manual for the AVSTAR.)

Now get your own AVSTAR out (or borrow one) and we'll do a run-through. Press the On/Clear key and check for a tiny triangle appearing in the extreme upper left corner of the display window. This means your batteries are in good condition. When the batteries start to go, the triangle will fade out of the display. The number 0 will appear when the computer is turned on. Anytime the limits of the computer are exceeded, the word "error" will appear in the display. For the first few tries, you will probably see this quite often—just remember: Garbage in, garbage out!

To clear the computer after an error has been made, press the On/C key twice. If you type in a wrong number, pressing the On/C key once clears the incorrect number without altering calculations already in progress. If an incorrect mathematical operation such as adding, subtracting, dividing or multiplication is made, press the On/C key and re-enter the problem.

The AVSTAR is set up to handle normal math operations much like any electronic calculator. The memory will hold stored data until the computer is turned off. To store a number, press the Sto key on the lower left. All previously stored data will be replaced with the number being stored. To recall this data to the display, press the Rcl key. Use of these two keys does not affect calculations in progress.

The AVSTAR is preprogrammed to make the following conversions: pounds/kilograms; feet/meters; nautical miles/kilometers; gallons/liters; Fahrenheit/Celsius; nautical miles/statute miles.

Let's try a conversion of the average human body temperature, 98.6 degrees Fahrenheit, into the Celsius scale. With the computer on and cleared, type in the 98.6 (don't forget that decimal), press the Convert key (lower left), and then the zero key. Notice that the blue fine print over that particular key indicates F to C, which is exactly the conversion we want. Your display should now read 37 (degrees Celsius), although your body temperature may be a little higher than that if you've had a few "error" displays by now. Read the fine print on the other keys to locate your other conversions, and try a few for practice (Quiz 3-1).

Now that you've mastered the conversions, let's launch into the navigation part. The keys for these operations are all inside the block outlined by colored pinstrips with the words "Nav Mode" to the left. There are three modes, all selected by one of the left column of keys: (1) Altitude/Airspeed (2) Wind and (3) Time-Speed-Distance.

The Alt/AS key engages the altitude/airspeed functions, turns on an A in the left part of the display window, and allows you to enter data with any of the keys inside of the black pinstriped area: Pressure Altitude, Density Altitude, True Celsius, Calibrated Airspeed, True Airspeed, and Indicate Celsius.

The Wind mode is surrounded by the red pinstripe and includes Wind Direction, Wind Speed, True Airspeed, Course, Ground speed/Fuel per hour, and Heading. Check your display for a B when in the wind mode.

The T-S-D mode is a "blue" area with keys for Time, Distance/Fuel, and includes the GS/FPH key. Also, take note of the "T:" key in the left column immediately below the T-

	CONVERT	TO
a.	140 nautical miles (knots)	statute miles
b.	25 knots	mph
c.	77 statute miles	nautical miles
d.	176 mph	knots
e.	18 degrees F	C
f.	13 degrees C	F
g.	36 gallons	liters
h.	246 liters	gallons
i.	36.5pounds	kilograms
j.	1380 feet	meters
k.	3546 meters	feet
l.	346 nautical miles	kilometers
m.	148 knots	mph

Answers: a. 161, b. 29, c. 67, d. 153, e. −8, f. 55, g. 136, h. 65, i. 17, j. 421, k. 11634, l. 641, m. 170.

Quiz 3-1. Conversions.

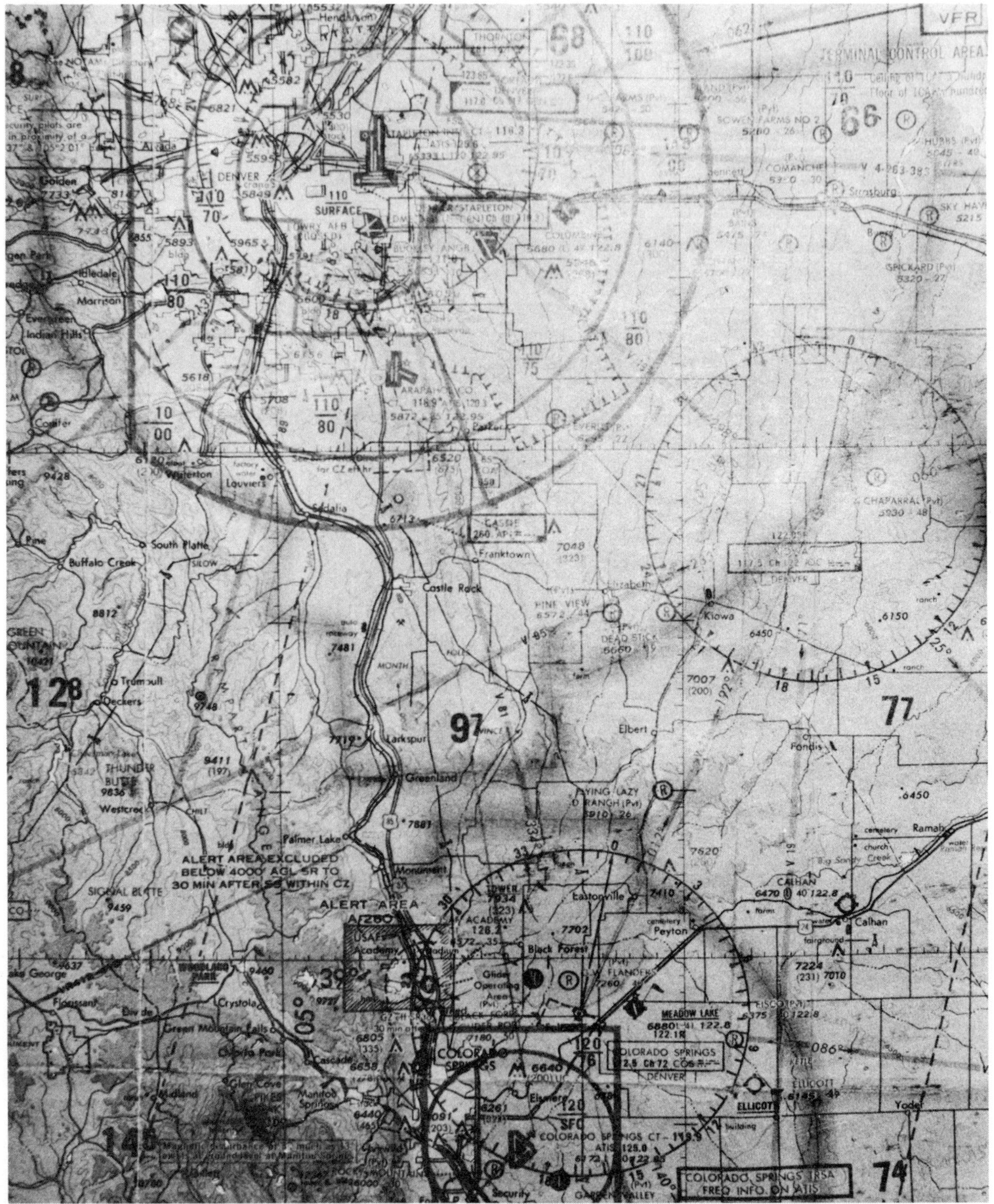

Fig. 3-2. Portion of Denver Sectional chart.

S-D key, which will be used in time computations, and the Compute key at the upper left, which will be used in all navigation problems.

We are now going to set up a simple cross-country flight to see how many of the above keys we can use. Follow me through carefully and don't be ashamed to ask your ground instructor for help; this first exposure will seem pretty long and involved.

THE SECTIONAL CHART

Now get out your Denver Sectional chart (Fig. 3-2) and let's get familiar with the legend section. This will be the same as the legend of all Sectional Charts. Notice first how the legend section (Fig. 3-3) is arranged as this will help you look up symbols more quickly, and the name of the game as you learn charts is to find answers on the legend. Also, don't miss out on other important information scattered elsewhere around the chart. The charts are updated every six months. Now try Quiz 3-2.

CROSS-COUNTRY PLANNING

Now let's use the Denver Sectional Chart (make sure you have the Sectional and not the Denver Terminal Area Chart—it shows a smaller area) and start planning a short trip to step through the preflight planning. We are going to fly from Arapahoe County airport (south edge of Denver) to the Kiowa Vortac as shown in Fig. 3-4. (The Vortac is the symbol in the middle of the compass rose.)

First, on your Sectional, draw a pencil line from the center of Arapahoe airport to the Vortac. This line will represent our True Course, which is simply the track over the ground that we wish to follow (Fig. 3-4). Determining the direction of our course requires a couple of simple steps starting with the use of your plotter as shown in Fig. 3-5. Notice that the small hole in the center of the plotter is always placed over the intersection of the course line and a longitude (N-S) or latitude (E-W) line. In this case, we used a longitude line which then becomes the pointer to show that our True Course (measured from the True North Pole) is 110 degrees. Prove this on your own plotter, and notice that if we were flying from Kiowa to Arapahoe, our True Course would be 290 degrees.

Now, look on the chart for a red dashed line, just to the west of Arapahoe Airport, that runs off toward the southwest. If you follow

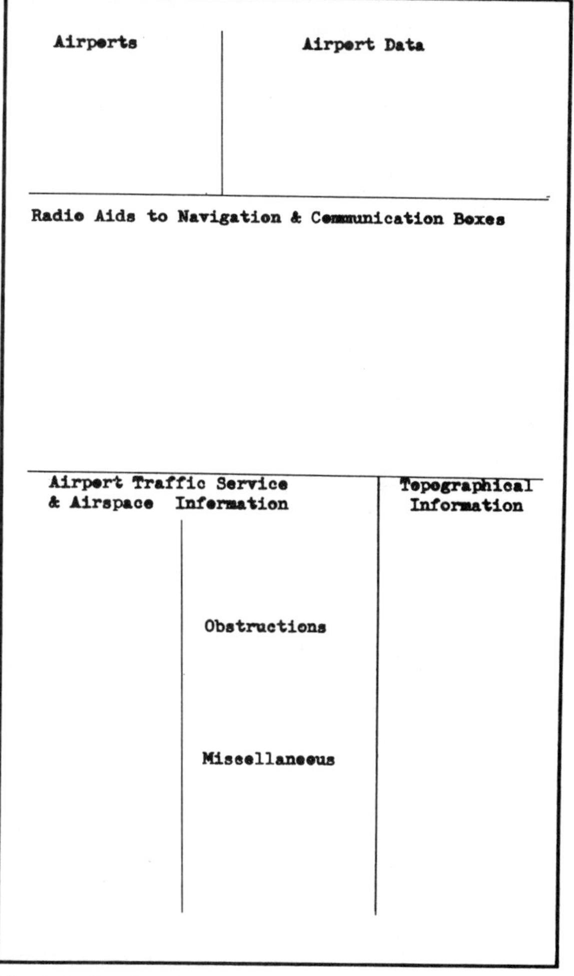

Fig. 3-3. Sectional legend.

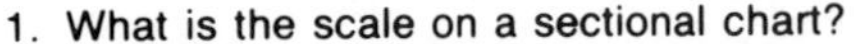

1. What is the scale on a sectional chart?
2. Charts may be updated by reference to:
3. What must a pilot do before entering a Restricted Area?
4. How about a MOA?
5. Information on these areas can be obtained where?
6. A Control Zone in which Special VFR is prohibited is shown by:
7. What is the symbol for a state line?
8. What is the symbol for a time zone?
9. What's the difference between a blue airport and a magenta airport?
10. Identify a military airport.
11. How can you determine if an airport has services?
12. Find the three different meanings indicated by tiny start: (IMPORTANT NOTICE—controlled airspace means weather minimums).
13. How are the limits of controlled airspace shown? What do the color codes mean?
14. What if controlled airspace floors are other than 700′ or 12′ AGL?
15. What is the floor of controlled airspace in a control zone?
16. Where is the base of controlled airspace over Weld County?
17. Where is the base of controlled airspace over Kugel-Strong (pvt) Airport south of Greeley, Colorado?
18. What are the field elevations for Weld County Airport and Cheyenne Airport?
19. Measure the longitude-latitude of Weld County Airport.
20. Here are a few more symbols you should be able to find.

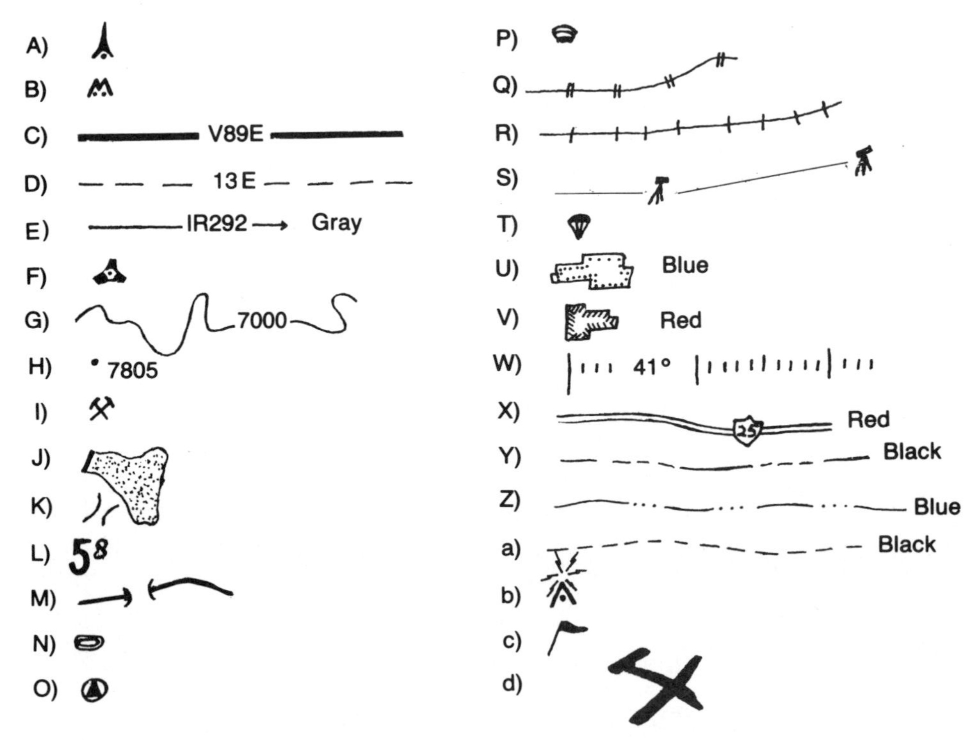

Quiz 3-2. Sectional legend.

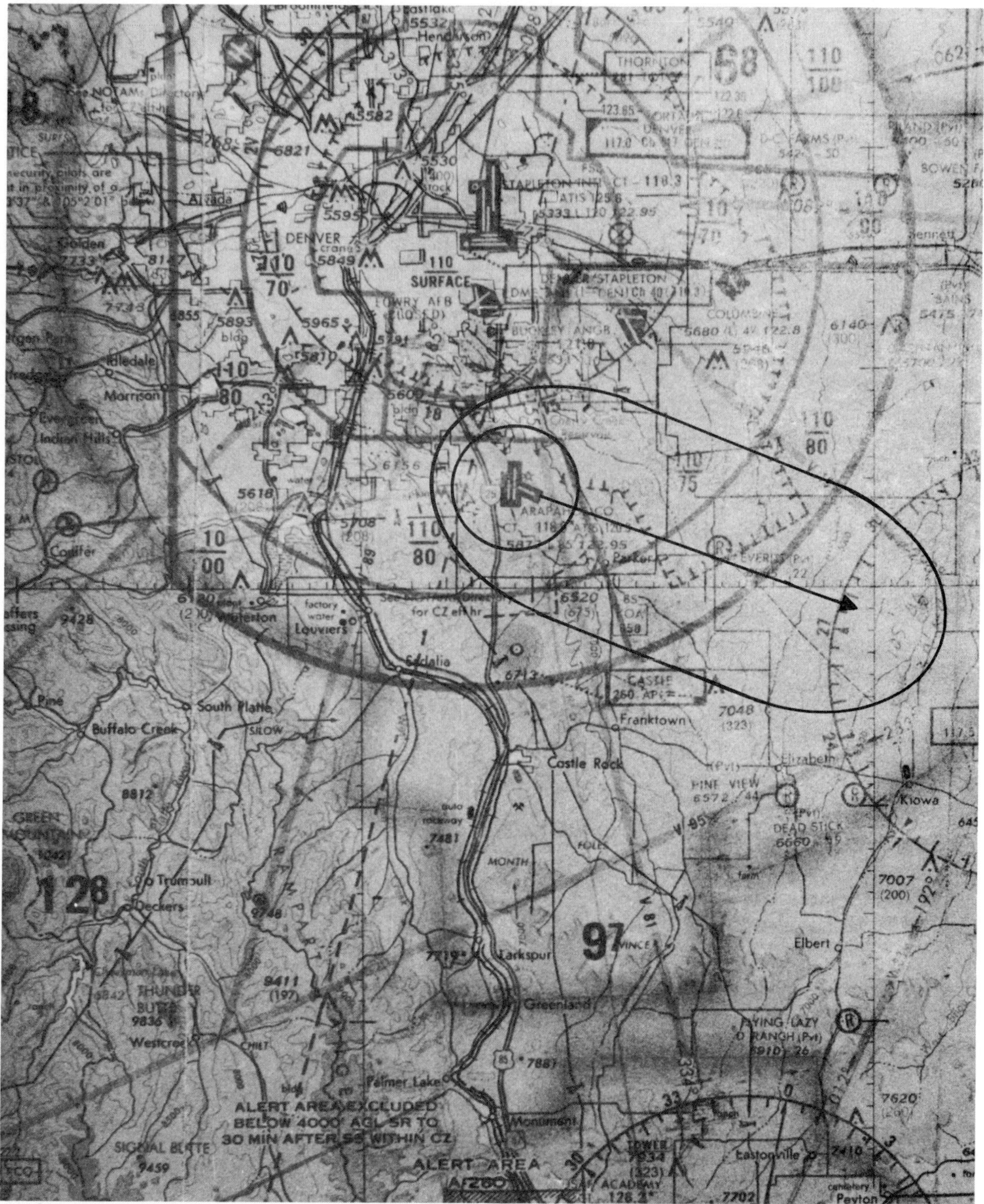

Fig. 3-4. Denver area.

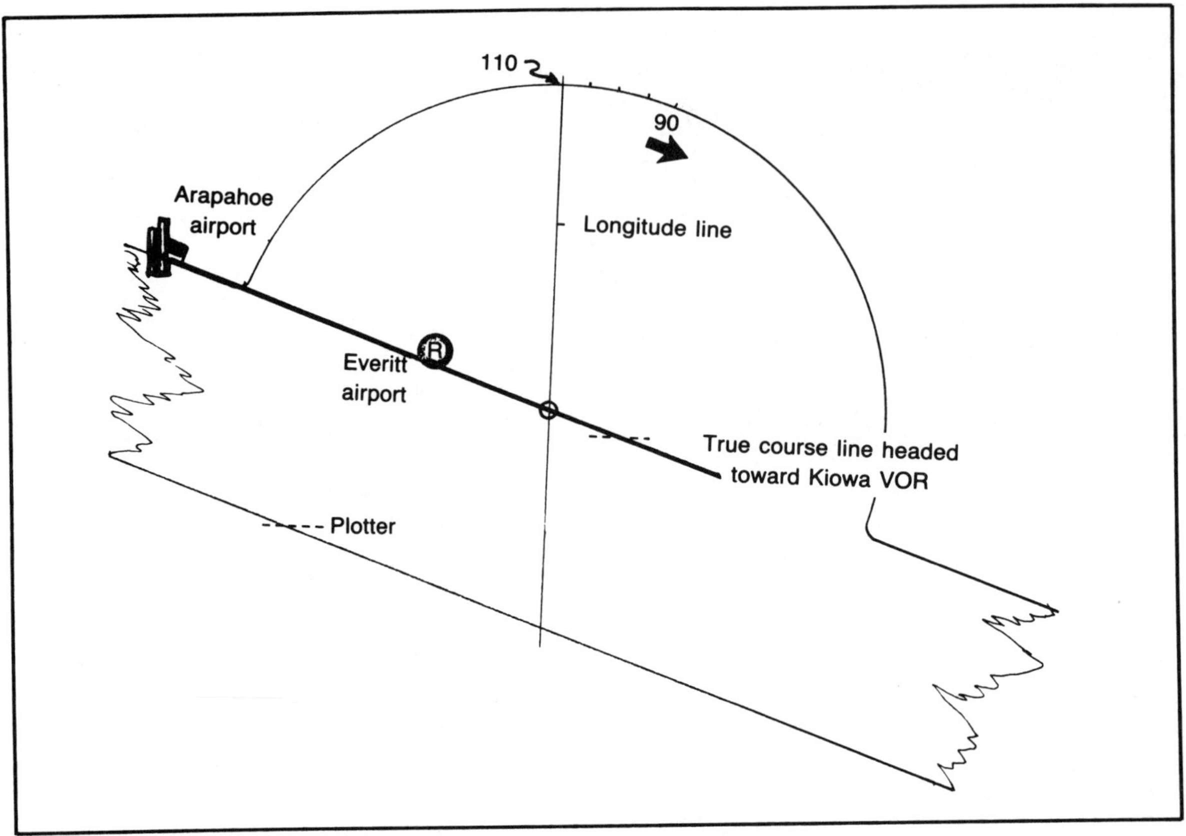

Fig. 3-5. Plotter and true course.

it down you will find the notation "12 degrees E." This is an Isogonic Line of Magnetic Variation (Fig. 3-6) and signifies that at all points along the line, a magnetic compass will be pointing exactly 12 degrees east of the True North Pole. This means that our True Course of 110 degrees must be adjusted to a Magnetic Course of 98 degrees (110 – 12 = 98). All "east" variations are subtracted, while all "west" variations (and they're all east of the Mississippi) are added. Pilots say "East is least and west is best" to help remember the procedure.

As you can see, we're getting into a lot of numbers fast, so let's get some organization going. One key to successful navigation is to organize all your information into a navigation log such as the one shown in Fig. 3-7.

THE NAV LOG

I've plugged in the figures for TC and VAR on our first leg. We don't really need to record the MC, but notice in the far right column is a reminder that we must select our cruising altitudes based on the direction of our Magnetic Course. In other words, let's break off and check out another FAA Regulation: FAR 91.109, "VFR CRUISING ALTITUDES," requires that we select an *even* cruising altitude plus 500 feet when our Magnetic Course is 180 through 359 degrees, and an *odd* cruising altitude plus 500 feet when our Mag Course is 0 through 179 degrees. This creates

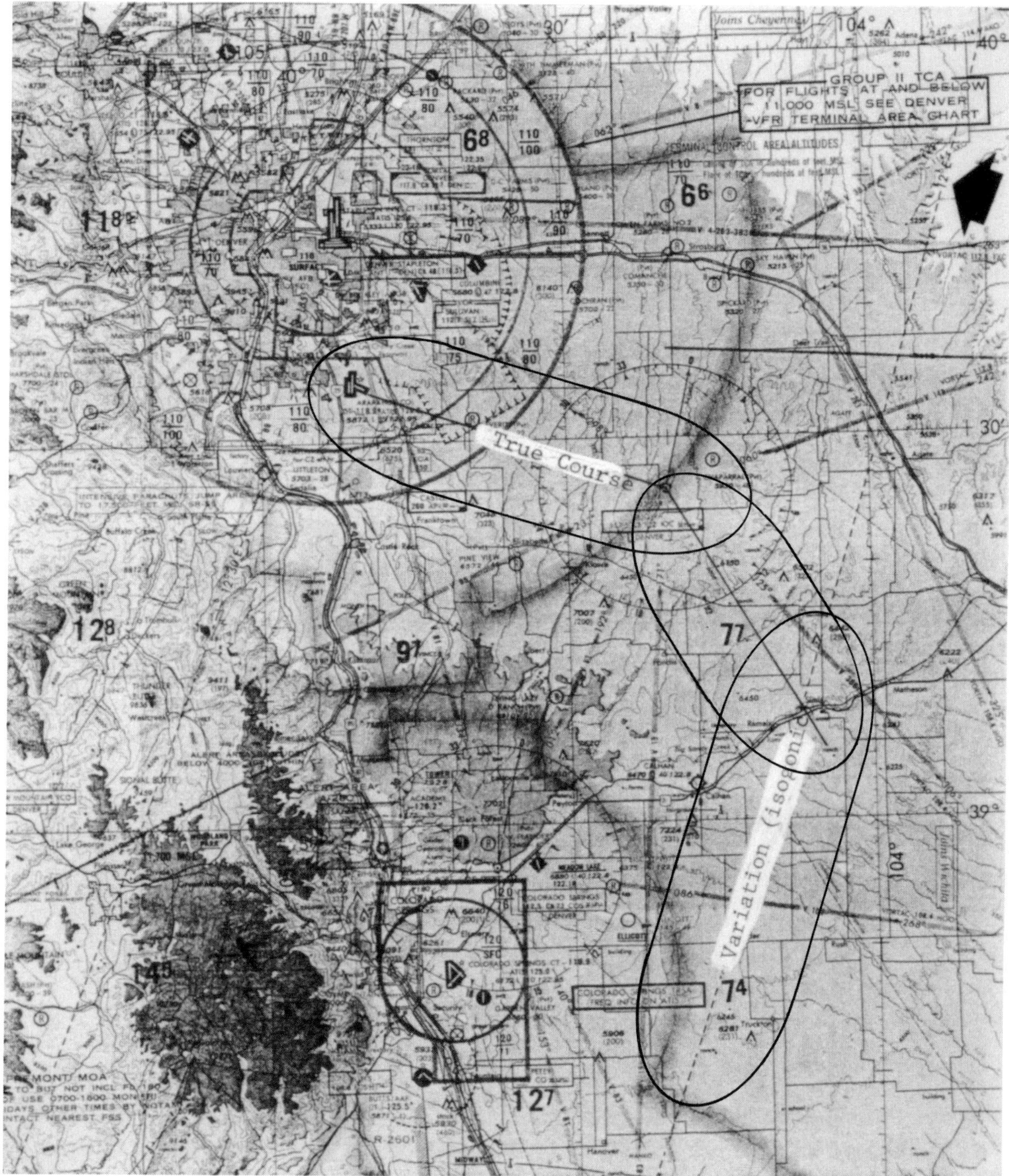

Fig. 3-6. Nav pointers.

a "layered" arrangement of traffic where the eastbound traffic flies at 5,500, 7,500, 9,500, etc., while the 6,500, 8,500, 10,500, etc. levels are reserved for westbound traffic. Our memory aid for this is "East is odd, west is even" . . . unless you'd prefer "Odd men fly east." Also, connect the MCs together and remember that Mag Course altitudes prevent midair collisions! Now check out the terrain and obstacles along the route and enter an altitude in the NAV Log.

What about the distance using the plotter? Make sure you are reading the correct side of the plotter. There are two sides to this gadget; one is for Sectional charts, and the other is for World Aeronautical Charts (which are half the scale of the Sectional). Each side has both a statute and a nautical scale. For this problem, let's stick with nautical, so you measure the distance from Arapahoe to Kiowa while I get us a weather briefing.

I've plugged in a typical true Airspeed for the Cessna 152, plus wind direction and velocity (in knots). See Fig. 3-8.

Now back to the AVSTAR to compute our ground speed and heading following the directions in the NAVSTAR instruction book. The difference between TC and TH will be your wind correction angle, which you should fill in as R or L for a right or left crab.

The next part of the nav log is the time and fuel consumption problem. Let's assume a fuel consumption of six gallons per hour—typical for the Cessna 152. Our TAS and fuel consumption will be measured on this trip so we can be more accurate next time.

The T-S-D (time, speed, distance) mode is probably the most used mode on the computer. We will use it now to estimate the time and fuel burn enroute, and use it again in flight to determine actual ground speed and revised estimates of arrival time (ETAs).

Follow me through the time, speed, and distance computations here:

Emergency	121.5
Emery Aviation	123.3
Flight Watch	122.0

	TAS	GPH		INSTRUCTOR'S SIGNATURE

LEG	TC	WIND	WCA	TH	GS	DIST	TIME	FUEL	VAR	MH	DEV	CH	ALT (MC)
1	110								12E				

CHECKPOINT	LEG	DISTANCE PT. CUM.	CLOCK TIME EST ACT	ELAPSED TIME	GROUND SPEED	

Fig. 3-7. Navigation log.

				TAS 90		GPH 6			INSTRUCTOR'S SIGNATURE				
LEG	TC	WIND	WCA	TH	GS	DIST	TIME	FUEL	VAR	MH	DEV	CH	ALT (MC)
1	110	060/10							12E				

Fig. 3-8. Nav log with wind.

1. Press T-S-D in the block with the blue pin stripe.
2. Key in our distance of 26 nautical miles.
3. Press Dst/Fuel.
4. Enter a ground speed of 83 kts and press GS/FPH.
5. Now press the Comp button to enter all data into the computation mode.
6. Press Time and read about 19 minutes.

This estimated time enroute is now written in the flight log in the time column.

Now, solving for fuel consumption:

1. Still in the T-S-D mode?
2. Press 00, then press T: (for hours).
3. Press 19, then press T: (for minutes).
4. Press Time, and all of that just entered the 19 minutes between Arapahoe and Kiowa.
5. Press 6 for the fuel consumption and then GS/FPH.
6. Now hit the Comp and then the Dst/Fuel for the 1.9 gallons of fuel out to Kiowa.

Finally, for our first leg, log in all of the data that we have computed in our problem. The TH is adjusted for VAR and entered as Magnetic Heading (MH). Deviation is the error of the compass for each individual aircraft. We can adjust the MH for DEV to get Compass Heading (CH) when we get out to the airplane (Fig. 3-9).

				TAS 90		GPH 6			INSTRUCTOR'S SIGNATURE				
LEG	TC	WIND	WCA	TH	GS	DIST	TIME	FUEL	VAR	MH	DEV	CH	ALT (MC)
1	110	060/10	5L	105	83	26	19	1.9	12E	93			7500

Fig. 3-9. Completed nav log.

In the lower part of the nav log, we will list visual checkpoints along the route of flight that we anticipate being able to see in flight. These checkpoints serve the dual purpose of checking to see that we're on course and also allowing us to time the distance between checkpoints for computing actual ground speeds and revised ETAs. (Remember that the winds were forecast only and we'll be flying in the real world!)

The AVSTAR handles this inflight problem easily:

1. T-S-D mode.
2. Enter distance between checkpoints (Dst/Fuel key).
3. Enter T: (hours), T: (minutes), T: (seconds), Time.
4. Press Comp and then GS/FPH for an actual ground speed.
5. Enter distance to the next checkpoint (DST/FPH), and Comp Time to the next checkpoint for the new ETA.

So you've now planned a short trip. At this time we would file a flight plan in accordance with our estimations. Since flight plans are not covered on the written test, I'll let your flight and ground instructors check you out on that part. Had it been a much longer flight, the nav log would have been a series of problems just like the above—one set of problems for each leg or course of the trip. Whether it's one leg or 14, the routine is the same:

1. Draw Course lines on the chart and measure and log the True Courses for each leg.
2. Measure and log the distances for each leg.
3. Choose cruising altitudes for each leg in accordance with MC and the cruising altitude regulation.
4. Compute True Heading and estimate ground speed for each leg in accordance with the forecast winds.
5. Compute time and fuel for each leg and for the total trip. Make your decisions about anticipated refuel stops at this time.
6. File a flight plan in accordance with the above estimates, and launch!
7. Enroute, fly headings that will maintain the drawn track across chart, timing yourself between previously selected checkpoints.
8. Recompute your ground speed more or less continuously, thereby ensuring that if the winds aloft should change, you'll be the first to know.
9. Recompute your fuel religiously. In this game you won't pull off on the shoulder and walk to the next gas station.

NAV PROBLEMS USING THE ELECTRONIC COMPUTER

Finding TAS

First, some definitions: IAS (*indicated airspeed*) is what the instrument is reading—regardless of altitude, temperature, or mechanical errors. CAS (*calibrated airspeed*) is IAS corrected for mechanical errors of the instrument and system. TAS is either of the above airspeeds corrected for altitude (pressure) and temperature. The best accuracy is by correcting IAS to CAS and then to TAS—the actual speed of the aircraft through the air.

Now try Quiz 3-3 on finding true airspeed.

Density Altitude

The *density altitude* is the most accurate altitude for measurement of aircraft perfor-

PROBLEMS:

	Assigned Altitude	Outside Temp	IAS	TAS
1.	3,000	25 C.	100	
2.	7,000	5 C.	130	
3.	4,000	15 C.	150	
4.	8,000	−5 C.	125	
5.	3,000	0 C.	95	
6.	5,000	0 C.	123	
7.	6,000	8 C.	145	
8.	8,500	20 C.	130	
9.	5,250	45 F.	160	
10.	4,000	26 F.	136	

Quiz 3-3. Finding true airspeed.

mance. It is readily found by running the temperature and pressure altitude through your computer. But first, we can figure the pressure altitude from knowing our field elevation and the current altimeter setting. You must remember that the sea-level standard pressure is 29.92 inches of mercury, and everything else is easy.

First get the difference between 29.92 and the altimeter setting. An inch of mercury is the equivalent of 1,000 feet of pressure altitude, so each 1/10 of an inch difference is 100 feet and each 1/100 inch pressure difference is 10 feet. Say the tower gives us a setting of 30.12:

	30.12
Minus	29.92
Difference	.20

which means 200 feet can be subtracted from field elevation to get pressure altitude. The subtract is because the pressure is higher than standard; that means more dense air, like you find at lower altitudes—so subtract. A low pressure day, below 29.92, will require adding the difference for a pressure altitude that is above field elevation because of less dense air.

So this set of problems is more accurate since we're working with pressure altitude (and CAS) to get density altitude and TAS out of the computer—like the engineers do it! Try Quiz 3-4.

Always remember that density altitude is a performance altitude! Since it's corrected for nonstandard atmospheric conditions, your aircraft will respond as if it was at that altitude on a standard (average) day. Service ceiling on a Cessna 150 is 12,650 feet above sea level on a standard day, but the actual point in space where that will be can vary quite a lot. The density altitudes at Weld County can vary from field elevation by plus or minus 2500 feet or more. At lower field elevations, on a cold day, the density altitude can actually go below sea level! On the other hand, when it's high altitude and hot days, it is possible for an aircraft to fail to become airborne. *Think density altitude!*

Finding True Heading, Magnetic Heading, and Ground Speed

As the wind becomes involved, you should

learn that the direction of the wind is always from (never to) and will be reported in directions from True North (except in airport advisories). Wind velocity is always in knots. And here's some more definitions:

True Course is a line drawn on a chart and measured as a direction in degrees from the True North Pole—not corrected for wind!

True Heading is the direction you will have to point the nose to keep the aircraft tracking the Course (still measured from true North). Your computer will solve for True Heading when programmed with Wind, True Course, and True Airspeed.

Magnetic Variation is the angle between the True North Pole and the Magnetic North Pole, as reported on the charts in the form of Isogonic Lines (lines of equal variation). From your True directions, subtract any variation shown as E for east, and add those shown as W or west. ("East is least and West is best.") The result will be a Magnetic Course or Magnetic Heading, as appropriate.

Magnetic Heading is the result of adjusting a True Heading for Magnetic Variation (assuming the wind was taken care of earlier—otherwise it would be a Magnetic Course).

Ground speed is the aircraft's speed measured across the ground. It's your TAS adjusted for the effects of wind. Try Quiz 3-5.

Finding ETE, Fuel Remaining, and Time Remaining

Because of the influence of wind, all fuel problems must be calculated in terms of "gallons per hour" instead of "miles per gallon." Now that you can compute your expected ground speeds, you're ready to try for a preflight estimate of the time required to complete the trip, the fuel burned during the trip, and the flight time remaining to you after you get there.

Now try Quiz 3-6.

RADIO NAVIGATION

VOR stands for Very high frequency Omni-directional Range-finder. A VORTAC is really the same thing but is set for joint civil/military use. Most aircraft today are equipped with a VOR receiver or NAVCOM, so this is the primary means of radio naviga-

PROBLEMS:

Field Elev.	Altimeter Setting	Pressure Altitude	Temp F or C	CAS	TAS	Density Altitude
4450	29.89		68 F	160		
6396	30.17		14 F	140		
2690	30.23		41 F	85		
7750	29.77		30 C	230		
3720	29.64		−13 F	110		
4212	30.10		86 F	400		
734	30.13		−20 C	140		
6240	28.88		50 F	105		
1420	29.84		0 C	223		
5510	29.31		15 C	90		

Quiz 3-4. Finding density altitude.

Direction/ Velocity	True Course	True A/S	Mag. Var.	True Mag. —Headings—		Ground Speed
020/10	090	140	10 E			
060/10	310	120	20 E			
330/35	148	178	0			
050/25	140	100	20 W			
300/10	175	206	9 E			
040/20	270	160	6 E			
270/20	112	151	10 W			
030/15	115	150	4 W			
150/10	090	161	6 W			
080/12	040	137	5 W			

Quiz 3-5. Finding ground speed.

tion in the country today. Frequencies range just below the communications band that you've already been using. Figure 3-10 shows the charted symbols for VORs.

Figure 3-11 shows the VOR as it looks on a sectional chart. I've put arrows pointing to the station location (top) and the azimuth or compass rose surrounding another VOR (bottom).

The information box near the station will tell you about the VOR and its monitoring facility (Flight Service Station) as follows (looking at KIOWA VORTAC): DENVER FSS will receive our transmissions on 122.05 MHz, and answer back on 117.5 MHz, the navigation side of the radio (be sure and turn up the volume). The "Ch" number is a channel for the military so we can ignore that. The IOC is a three letter identifier unique to KIOWA and we can listen to the Morse Code for IOC which follows anytime the Denver FSS isn't talking to us. If we fail to hear this identifier when tuned to Kiowa's frequency it means one of two things: If our VOR navigation is working, then some maintenance is being performed that may distort the nav signal—it should not be relied

Ground Speed	Total Dist.	Flight Time	Fuel per Hour	Useable Fuel	Fuel Left	Time Left
110	115		5	20		
120	310		9	26		
125	660		12	80		
170	814		11.5	60		
135	170		8.5	32		
245	1,470		36	160		
115	515		11	50		
85	230		4	18		

Quiz 3-6. Finding ETE, fuel used, and time left.

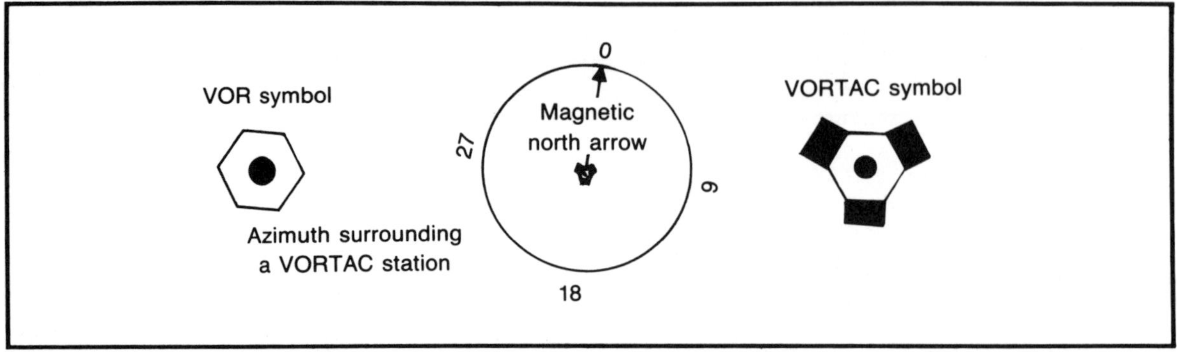

Fig. 3-10. VORs and VORTACs.

upon. Or if the VOR navigation equipment is *not* working, then consider climbing to a higher altitude or selecting a closer VOR. All aircraft frequencies designated by MHz (MegaHertz) are limited by what we call the "line-of-sight" distance—meaning that at lower altitudes, your nav and communications may be severely limited.

The VOR

The VOR station broadcasts a signal in all directions from the transmitter much like the spokes of a bicycle wheel, and each spoke is called a *radial.* These are measured in degrees from Magnetic North. This means that all navigation with the VOR will be based on magnetic directions, thus bypassing the corrections for Magnetic Variation that we were making earlier.

Now let's navigate using the VOR. Say you are wanting to go in the direction of the course (radial) shown in Fig. 3-12. To follow that radial you have to take two separate actions:

1. Tune the Colorado Springs VORTAC on the frequency of 112.5 MHz (black arrow) on the NAV side of your radio. Turn up the volume and confirm the COS audio in Morse Code to be sure you have the right station and that you are receiving a reliable signal (Fig. 3-13).

2. Turn the Omni Bearing Selector (OBS) or Course Selector (CRS) knob at the lower left of the VOR head until 120 appears at the top of the instrument. You have now selected the Magnetic Course desired for the flight (Fig. 3-14).

The electronic mechanism of the VOR is set up so that along a line perpendicular to the selected radial we have separation of the TO and FROM zones. If your aircraft is in the TO zone (when you select the 120 course on your OBS), the TO/FROM flag will flop to the TO setting. If you are in the FROM zone it will show a FROM. (If an unreliable signal is received or if your aircraft is located on the boundary between the TO and FROM zones—perpendicular to the selected radial—then the TO/FROM flag will show an OFF indication).

The To-From Zones

The small airplane shown in Fig. 3-15 is in the TO zone. Always remember that the exact placement of the TO and FROM zones is determined by nothing more than which radial you select on the OBS, *not* by whether the aircraft is flying toward or away from the station.

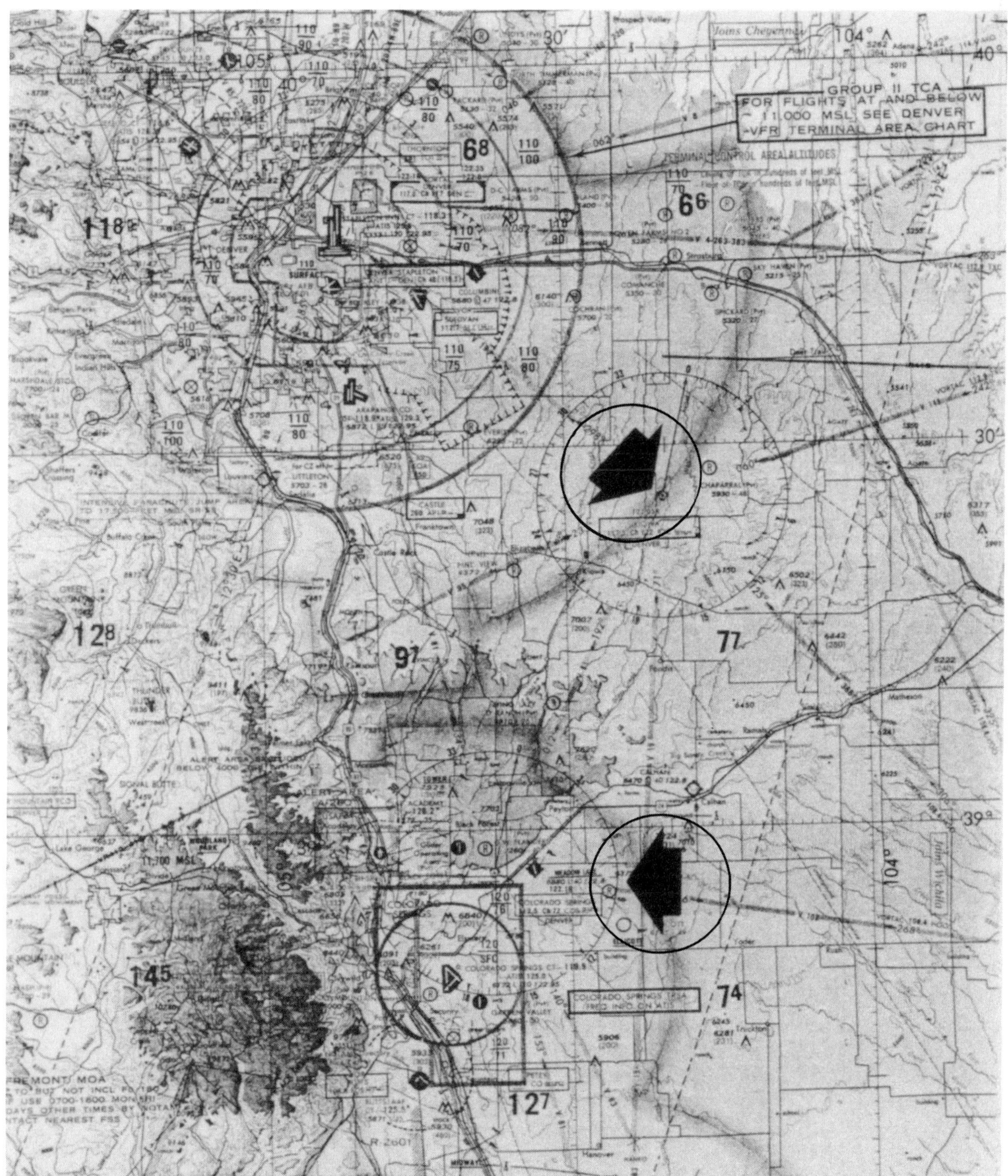

Fig. 3-11. VOR symbols and azimuth.

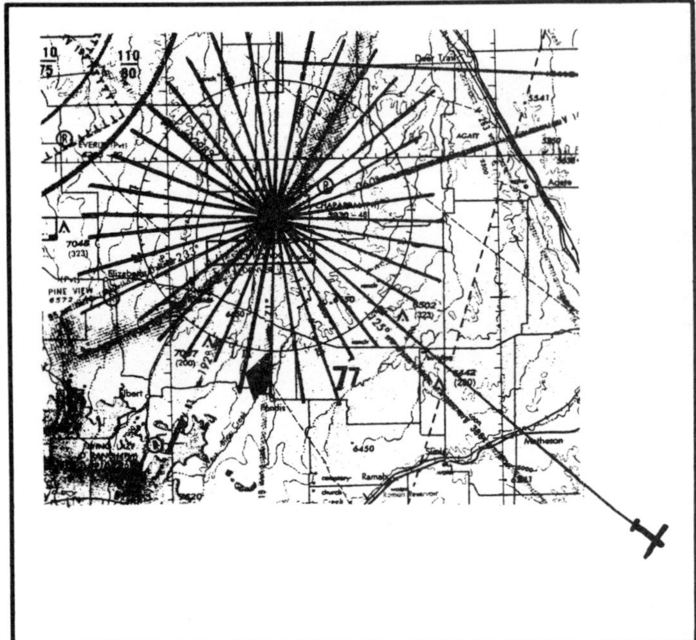

Fig. 3-12. VOR radials (10 degrees shown).

If that airplane turns around and flies away, he will still be reading a TO.

The VOR is electronic and can only sense a signal that is sent to it. Regardless of the aircraft's heading, the VOR only shows the zones that are relative just to your selection of a radial. This same principle extends to the "needle" or CDI (Course Deviation Indicator) shown in Fig. 3-16.

The Right and Left Areas

Figure 3-17 shows that the selected radial defines a boundary between a left area and a right area, along which your CDI needle will be centered. If I fly FROM the COS VORTAC on the 120 course, I will have a FROM flag and a centered needle. If the wind drifts me off course to the north (left area), the CDI will point to the right, telling me how to get back

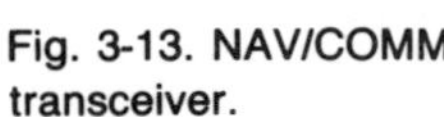

Fig. 3-13. NAV/COMM transceiver.

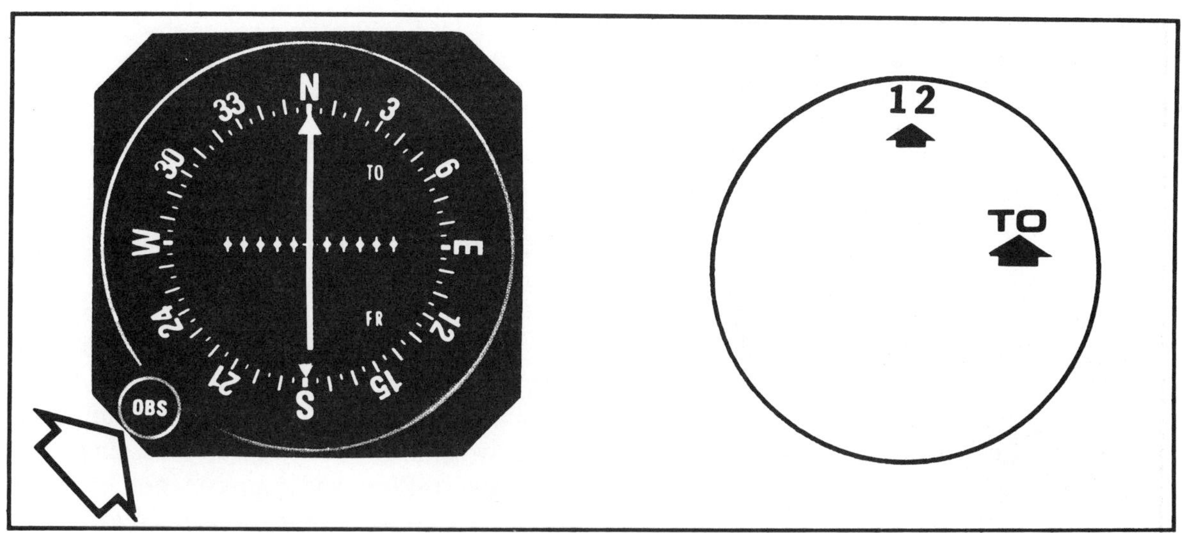

Fig. 3-14. VOR head and OBS.

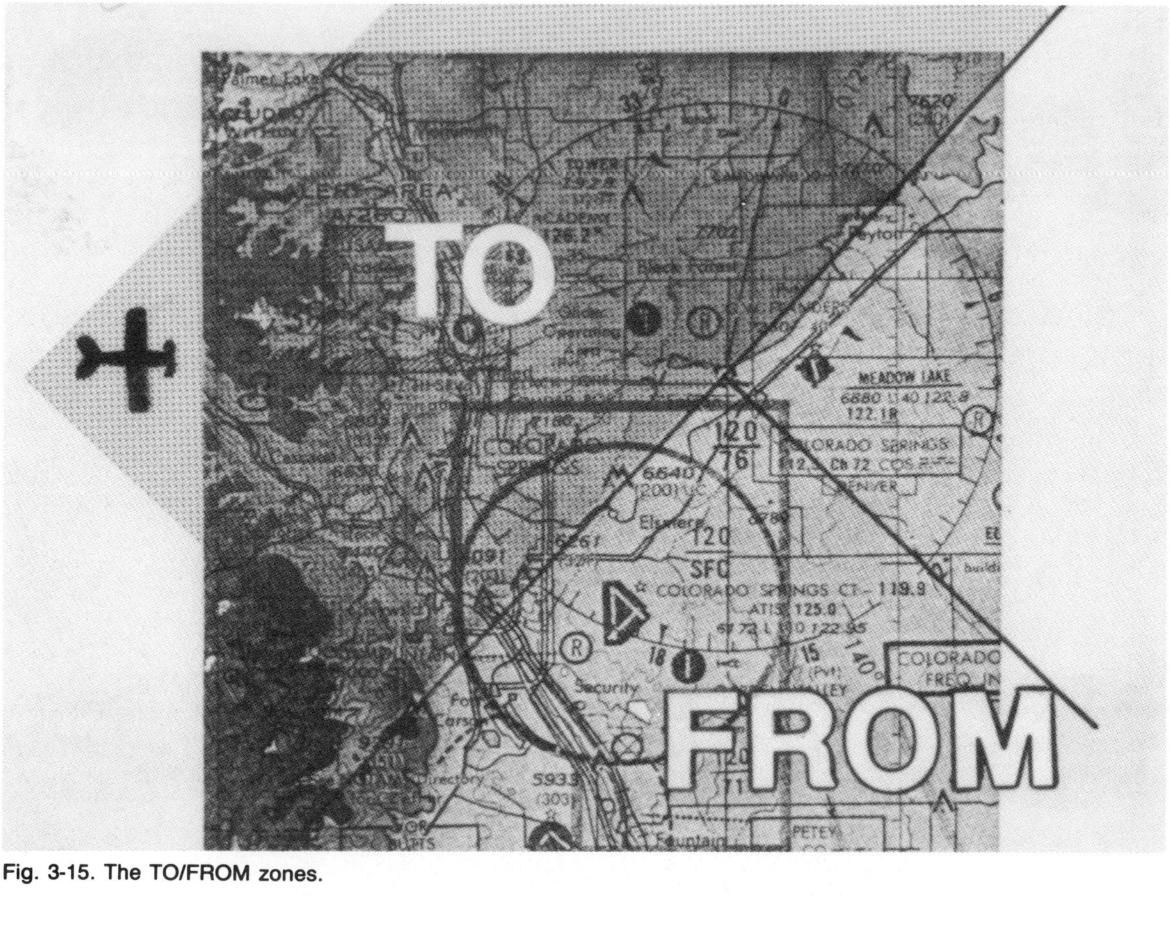

Fig. 3-15. The TO/FROM zones.

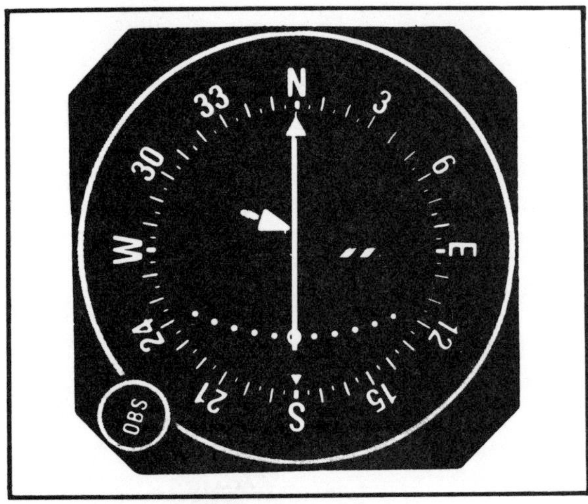

Fig. 3-16. The CDI.

on course. If I drift off course to the south, the CDI tells me to turn to the left (Fig. 3-18).

Practically Speaking

Anytime your aircraft is off course, the needle will sense the area and give you the appropriate indication. The best way to keep this straight is to always (mentally or otherwise) turn the aircraft so that the heading agrees with the OBS (radial) before reading the VOR! Then you will know that your instrument is properly sensing your location for flying to or from the station, and you can get back on course by simply "flying the needle." Only

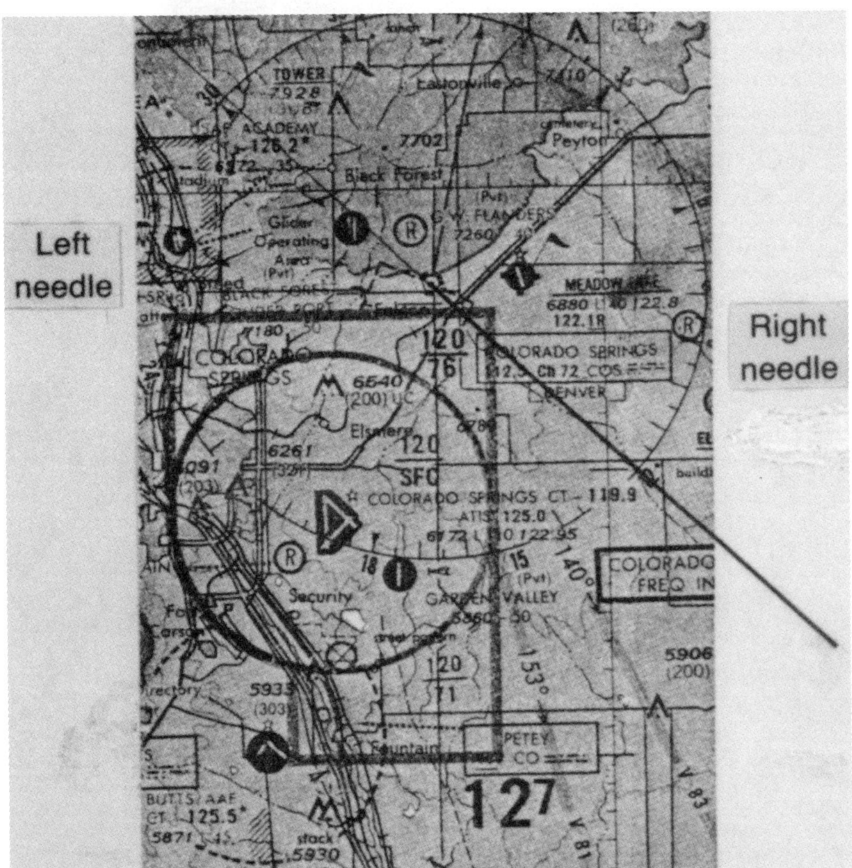

Fig. 3-17. CDI right and left areas.

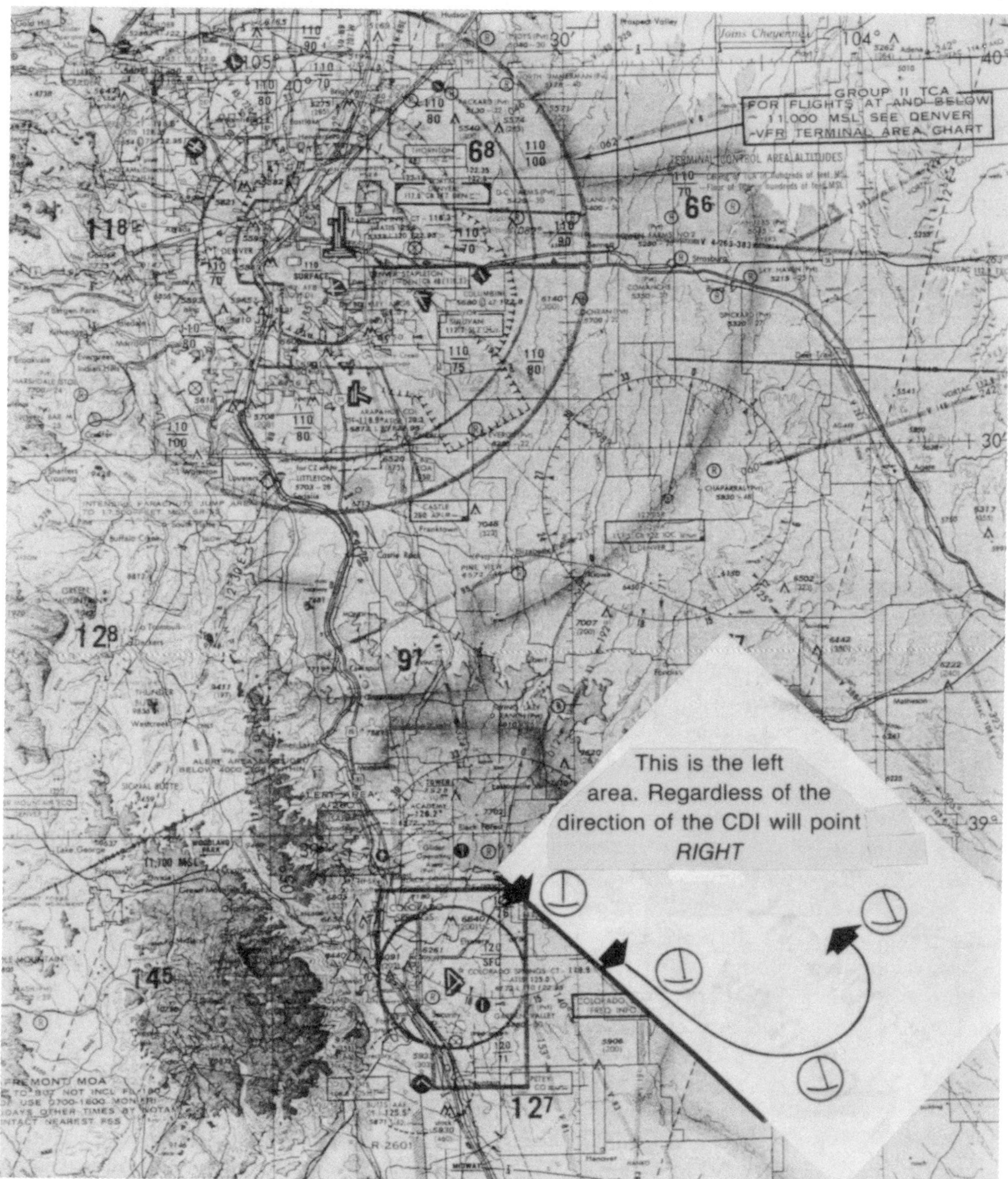

Fig. 3-18. Aircraft flying off course outbound on 120 radial COS.

when the heading and the OBS are in general agreement will this work.

Our primary use of the VOR is for straight-line navigation to and from the station itself. This means when passing over the station, there is a greater traffic hazard than ordinary and your eyes should be outside! Station passage will be signaled by the CDI becoming overly sensitive, the TO/FROM changing to an OFF for a few moments directly over the station, then changing to a FROM as we go out into the FROM zone. If our desired course does not continue in a straight line across the station, the new radial should be selected with the OBS at this time. But watch out for that traffic!

A secondary use of the VOR is to create "radio" checkpoints in addition to the visual checkpoints mentioned earlier. For example, choose radials (preferably in the FROM zone) that lie across (rather than along) your course. The time it takes to fly between those two radials (centered needle) can be programmed with the distance between them (chart and plotter) for a fairly accurate ground speed check.

A couple of points should be covered before going further. The Victor Airways, shown on the chart as blue lines running between the VORs, are heavily traveled routes. They are given numbers, such as V83 between Kiowa and COS, and they are sort of like highway numbers. More importantly, the radials to fly the airways are shown leaving each VOR. The airways are one of our types of controlled airspace, thereby requiring stricter VFR weather minimums.

A VOT, or VOR test facility, is located at some airports. This facility is basically a 360° radial that is transmitted in all directions so that when the frequency is tuned in the VOR, the OBS should show a 360 FROM or a 180 TO indication with a centered needle. This indication should occur at any point on the airport surface. These testing facilities are listed in the Airport/Facility Directory and are used to test the accuracy of a VOR receiver.

The Automatic Direction Finder

The Non-Directional Beacon is a ground-based navaid that is broadcast in the low-frequency range. NDBs are charted with a red (magenta) symbol that is easily found on your legend (Fig. 3-19). ADF stands for Automatic Direction Finder and is the airborne equipment which tunes to the NDBs—or, for a higher initial cost, to the local commercial broadcast stations. One of the advantages this navaid has over the VOR is that you're not limited to "line-of-sight" signals like the VHF frequencies are. The low-frequency signal can bounce off the ionosphere and travel great distances. The VOR signal, on the other hand, is a straight lien and if you happen to be on the other side of a mountain, or the Earth's curvature, you can't receive the VOR signal. It is somewhat like FM radio, whereas the NDB is more like AM radio.

The ADF instrument in the airplane always points to the station that is currently being received. An aircraft outline is depicted on the instrument so that we can locate the station "relative" to the nose of the airplane. If the needle points to your right wingtip, we say you have a *relative bearing* of 090 degrees—which means that a right 90 degree turn will point you at the station. Such a turn will, of course, change your Relative Bearing to 0 or 360.

But please note that all of the above tells you nothing about which direction you are going, or which heading you will fly to the station. Figure 3-20 shows an aircraft with a

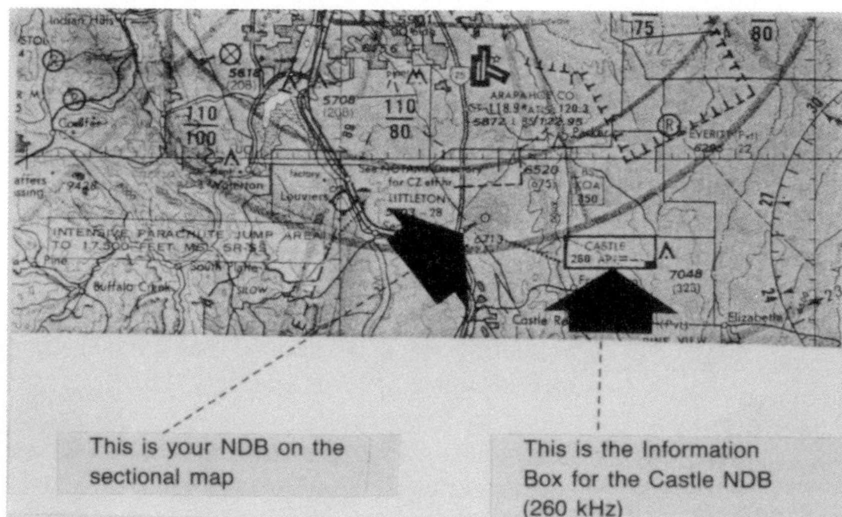

Fig. 3-19. NDB south of Denver.

Magnetic Heading of 008 degrees and a Relative Bearing of 055 degrees. His Magnetic Bearing to the station (the direction he must fly to go there) is 063 degrees. How did I do that? By turning right 55 degrees (RB) from a heading of 008 (MH). In effect it's a simple formula: MH + RB = MB to STA. For the FAA written test, remember that formula and just add your compass (or DG) heading to the ADF indication (Relative Bearing) for the Magnetic Bearing to the station. If you add the two numbers and the result is more than 360, that just means you're going around the compass rose the second time, so simply subtract

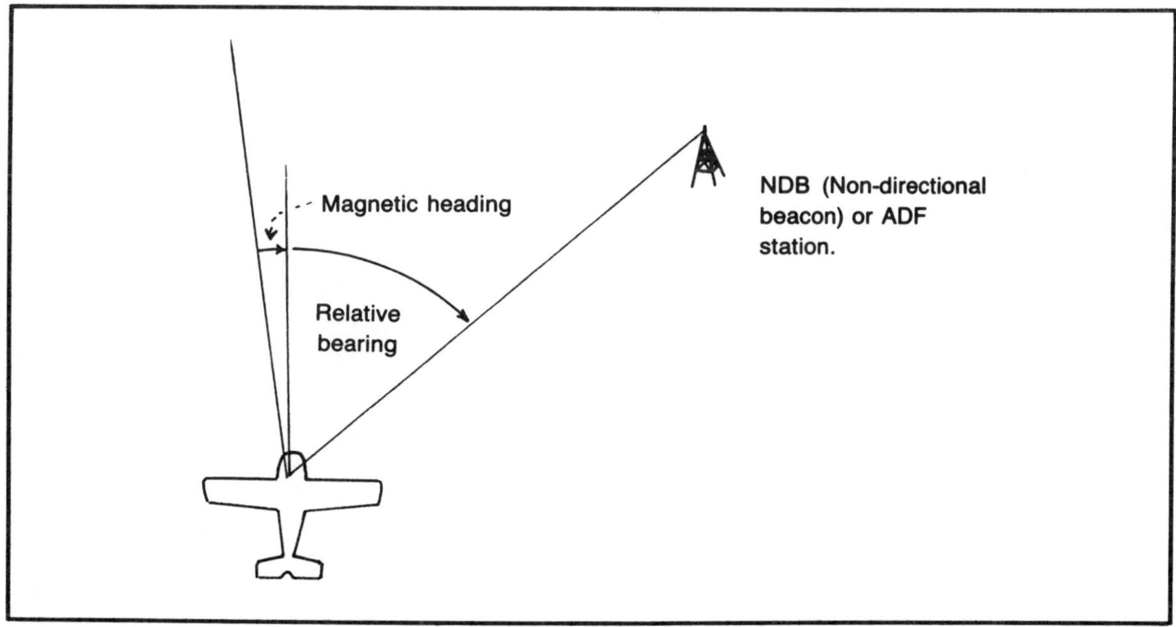

Fig. 3-20. ADF orientation.

360 and the result is your Mag Bearing to station.

An RMI, on the other hand, combines a DG with an ADF needle set on top of the compass rose. the important thing to remember is it does the calculating for you. To find MB to station, just look at the number the needle is pointing at. An outbound bearing, or bearing from the station, is found by looking at the tail of the needle. The magnetic heading is located at the top of the compass rose since this is actually a DG. The letters RMI, by the way, stand for Radio Magnetic Indicator.

AIRSPACE AND WEATHER MINIMUMS

Flying is primarily based on the pilot's ability to see and avoid other aircraft and obstacles. If the lack of visibility or clearance from clouds would prevent a pilot from seeing another aircraft in time to avoid it an unsafe condition would result. In *controlled airspace*, the visibility and distance from cloud requirements, called *VFR Minimums* (VFR stands for Visual Flight Rules), are increased from the uncontrolled airspace requirements to allow the VFR pilot to see and avoid other aircraft. If the conditions were below the VFR minimums, only IFR aircraft would be allowed to utilize the airspace. Since the pilot flying in IFR conditions cannot maintain visual separation from other aircraft, a ground-based controller provides separation by utilizing time, altitude, speed, or distance criteria.

Generally speaking, controlled airspace is airspace where the FAA imposes more strict weather minimums for the purpose of separating IFR and VFR aircraft when the conditions are becoming marginal. The VFR aircraft cannot be there (legally) and the IFR traffic will be separated by the controller. Regardless of the specific type of controlled airspace involved, it is the pilot-in-command's responsibility to maintain the VFR cloud separation and visibility requirements specified for that airspace. If that becomes impractical, or impossible, the pilot would have to maneuver the aircraft into uncontrolled airspace or obtain an IFR clearance.

Types of Controlled Airspace

The different types of controlled airspace are defined by geographical limits and altitude in FAR Part 71, summarized below:

Control Zone: Controlled airspace which extends upward from the surface of the earth. A control zone may include one or more airports and is normally a circular area with a radius of 5 sm, but often has extensions necessary to include instrument approach and departure paths. Marked on the chart as dashed blue lines. Designates an airport with a qualified weather observer. Basic VFR rules require 3 miles visibility and no ceiling below 1,000′ AGL. Special VFR clearances (FAR 91.107) are possible if the zone is below basic VFR minimums, down to 1 mile visibility and clear of clouds. Special VFR will not be granted to VFR pilots in a control zone marked by a "T" between sunset and sunrise, or when the visibility is less than 1 mile and, or, the pilot is not able to fly clear of clouds.

Transition Area: This area extends upward from 700′ AGL and designates an airport with an instrument approach procedure. Marked on the chart as a magenta area. Notice that you can fly below the transition area to obtain uncontrolled airspace weather minimums of 1 statute mile and be clear of all clouds.

Control Areas: These include all blue-shaded areas on your Sectional. Take a look and be sure you understand where the controlled airspace is from the shading. The blue is considered to start from 1200′ AGL and extend upwards, but often a nonstandard floor will be charted as an MSL base of the controlled airspace. Notice that all Victor Airways are included in the blue airspace.

Continental Control Area (CCA): Covers the airspace of the 48 states plus portions of Alaska, at or above 14,500′ MSL, but not including the airspace at or below 1500′ AGL. In other words, the floor of the CCA is recessed over the mountain peaks to 1500′ above the peak.

Positive Control Area (PCA): Starts at 18,000′ MSL and is the VFR pilot's upper limit. You are prohibited from flying above 18,000 feet (Flight Level 180) until you get an instrument rating.

Terminal Control Area (TCA): This is an area marked on charts by heavy blue lines with vertical limits listed as MSL values. These are located at very high density airports and are divided into two groups based on the size of the airport. There are certain restrictions to using the TCA so don't open or close flight plans or ask for weather briefings on this frequency even though the weather is their business.

Figure 3-21 is the FAA's attempt to graphically portray these airspaces and a few more coming up. Use your chart to get the "three-dimensional" picture . . . it's *very* important!

Those All-Important Minimums

This list is to help you separate the really important facts from the "nice to know" facts about our airspaces. The other notes you have on these spaces are handy to know, but the listing here will be the most important facts.

Controlled airspace means weather minimums! In other words, we want to know if it is controlled or uncontrolled airspace so we know which set of minimums (Fig. 3-22) apply to us.

What is considered controlled airspace? All of the following are:

- ☐ Control Areas—Blue shaded areas of the sectional charts and include Federal Airways. This airspace begins at 1200′ AGL unless noted by an MSL value, and continues upwards to but not including 14,500′ MSL.
- ☐ Transition Areas—Magenta shaded areas around airports on the sectional charts. Begins at 700′ AGL and goes up to the overlying controlled airspace.
- ☐ Control Zones—That area marked on charts by a blue dashed line 5 sm in radius, plus extensions, and up to the Continental Control Area. Special VFR must be obtained to enter this airspace when the Visibility is less than 3 miles, or the ceiling is less than 1000′. Remember that Special VFR may *not* be obtained when visibility is less than 1 mile, the pilot cannot stay clear of clouds, between sunset and sunrise, and when the control zone is marked by Ts.
- ☐ TCAs—Marked on the charts by a heavy blue line. The vertical limits are shown as MSL values. The requirements to enter any TCA are as follows:

  —Operable VOR.
  —Two-way radio.
  —Transponder.
  —Clearance.

Group I TCAs, add the following:

—Mode C transponder.
—Private Pilot, or better, to land or take off at airport within TCA.

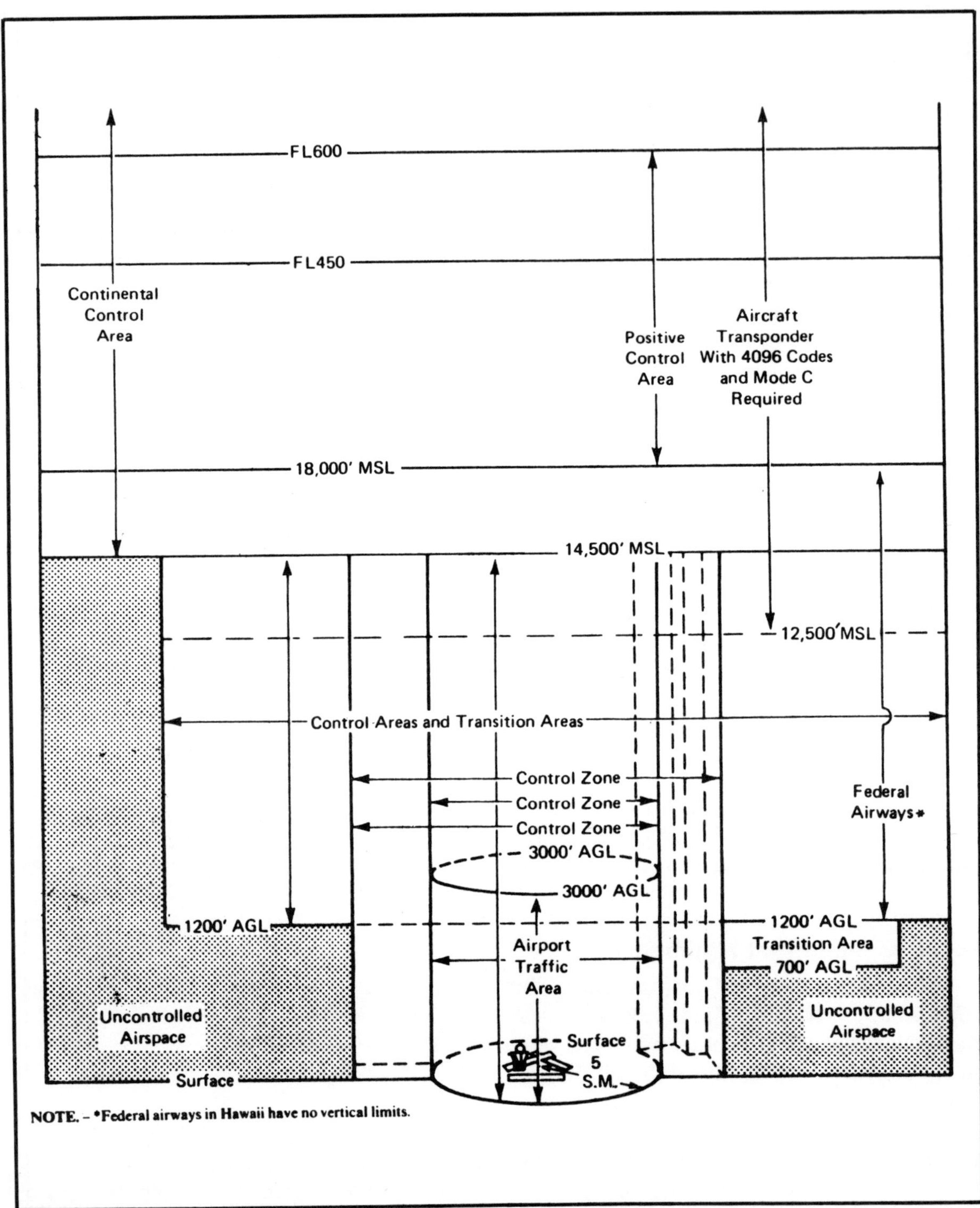

Fig. 3-21. Airspace graphic display.

	CONTROLLED	UNCONTROLLED
At or below 1200′ AGL	3 miles visibility 500′ below 1000′ above 200′ horizontal from Clouds	1 mile visibility Clear of all clouds
Above 1200′ AGL but less than 10,000′ MSL	3 miles visibility 500′ below 1000′ above 2000′ horizontal from clouds	1 mile visibility 500′ below 1000′ above 2000′ horizontal from clouds
At or above 10,000′ MSL and above 1200′ AGL	5 miles visibility 1000′ below 1000′ above 1 miles horizontal from clouds	5 miles visibility 1000′ below 1000′ above 1 mile horizontal from clouds

Fig. 3-22. VFR minimums for visibility and distances from clouds.

- ☐ Continental Control Area—Overlies the conterminous 48 states and parts of Alaska. At and above 14,500 MSL when above 1500 AGL.
- ☐ Positive Control Area—At and above 18,000 MSL. This is the VFR pilot's upper limit.

Airport Traffic Control

Airport Traffic Area (ATA): This one is also shown in Fig. 3-23, but please remember that it doesn't relate to weather minimums. It's just that control towers seem to always be sitting in a control zone (although plenty of control zones do not have a tower!). The tower does not change my weather minimums, but when the tower is operating, you must establish two-way radio contact when flying within 5 statute miles of the airport, up to but not including 3000′ AGL. (The only exception is when operating directly to or from a secondary airport located within the primary airport's Airport Traffic Area.) FAR 91.87 tells you to follow the tower's directions, and FAR 91.77 allows the tower to communicate by means of light gun signals if your equipment fails. (A few towers are designated on the

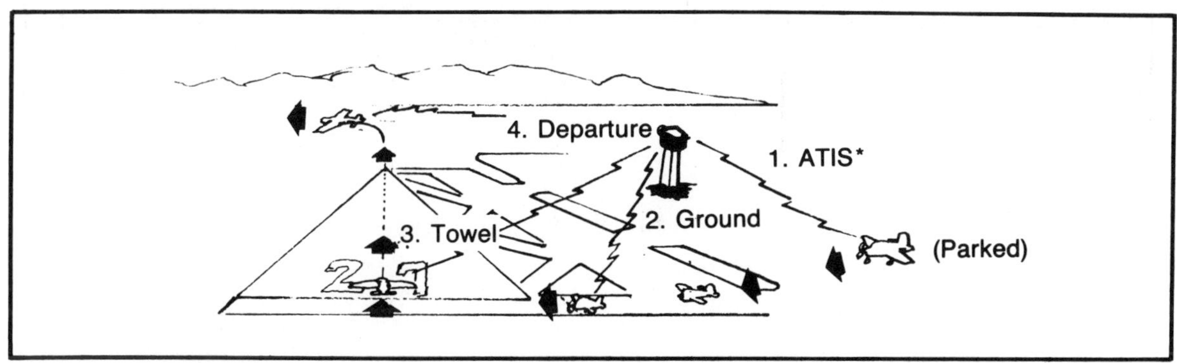

Fig. 3-23. Airport communications.

charts by "NFCT" for Non-Federal Control Tower. These are operated by private companies, but all the same rules apply.)

Terminal Radar Service Area (TRSA): Designated by a thick red outline on the chart (Colorado Springs is our local example) with floors and altitudes shown. The TRSA offers sequencing and separation services when designated as a Stage III, and a Stage II offers advisories and sequencing. These services are provided by the radar controller by assigning altitudes, headings, and possibly the location of nearby traffic, much like the controller in a TCA. But the difference is that your participation in the TRSA program is not mandatory. (Of course, you will communicate with the control tower before the ATA).

Stage II Radar: Is not designated on the chart, but check the Airport Directories for Pueblo, CO, or Rapid City, SD, for local examples. Here the radar controller offers Traffic Advisories and sequencing for participating VFR aircraft.

Basic Radar Service: Offers Traffic Advisories and limited vectoring to VFR aircraft.

Airport Communications

Let's take a close look at the sequence of communications when departing an airport that has the full array of communication facilities. Figure 3-23 shows what is probably a TCA or TRSA but could certainly be a simple Airport Traffic Areas.

1. Automatic Terminal Information Service, or ATIS, is a tape recording of non-traffic control information used at high activity terminal areas. Parked on the ramp we can tune the ATIS frequency shown on the chart (check Denver) and hear something like, "This is Stapleton Airport, information Bravo, measured ceiling 5,000 broken, visibility 12, wind 260 at 8, current altimeter 29.90. Landing runway 27, takeoffs 36. Approach Control 119.6, Ground Control 121.9. Advise you have information Bravo." After that plays over a few times, you have all the numbers! That "Bravo" means that as soon as anything changes (such as a wind shift) they'll change the tape and call it information "Charlie."

2. Ground Control will clear you to taxi wherever you need. Here's one for the FAA written test: If Ground Control clears you "to Runway 36," that "to" means you can cross over any runways between you and 36, but not to taxi "on" 36.

3. Control Tower must be contacted when you're ready for takeoff. After you're airborne, he will "hand you off" to:

4. Departure Control, who will channel you away from the airport without conflicting with other traffic.

When landing, the sequence is reversed, except that ATIS is always first, so it's like this: ATIS, Approach Control, who hands you off to Control Tower, who will instruct you when to contact Ground Control. Easy!

Radar And Transponders

The 4096 Code transponder is an airborne radio that sends signals to a radar set on the ground. You can dial in a four-digit code (1200 is standard code for VFR, unless otherwise directed), which appears on the controller's radar screen and follows your aircraft along its flight path. If the controller advises you to "Squawk 4321," then you just dial it into your transponder. If he also tells you to "Squawk ident," you push an "ID" button on your transponder, which makes your blip light up on his screen for positive identification. As you leave his area, he will tell you "Radar service terminated . . . squawk VFR" and you return to the 1200 code. Learn the following transponder codes too, and be careful not to "switch through" the 7000 series:

- ☐ 7500 means hijacked!
- ☐ 7600 is lost communications.
- ☐ 7700 is emergency.
- ☐ 0000 is for military intercept use *only*.

If you don't have a transponder, on your initial call-up he'll probably have you make a turn or two to see which blip moves, and then advise "Radar contact, 5 miles east of Litterbarrel" or wherever. Anytime you're working with radar, be ready to hear him say "Traffic at 2 o'clock," or whatever. It's easy! Just picture your aircraft on the face of a clock. Your nose is 12, right wing 3, left wing 9, and the tail is 6 o'clock. You might keep in mind, though, that he's seeing your *track* instead of your *heading*, so your "clock" might get shifted by your wind correction angle.

Transponders come in two varieties: Mode A and Mode C. The Mode C not only sends the controller your squawk code but is also connected to your *encoding altimeter* to send the controller your altitude as well. Mode C is required when operating in Group I TCAs, and when operating in controlled airspace above 12,500′ MSL, unless below 2500′ AGL.

The Flight Service Station

Although not an ATC (Air Traffic Control) facility, the Flight Service Station (FSS) does give advisory service if they have an office on a non-tower controlled airport. This is a more dependable traffic advisory than working with Unicom and will be done on a standard frequency of 123.6 (shown above the FSS box). This frequency is reserved for airport advisories and traffic use only, so don't open or close a flight plan or ask for a weather briefing on this frequency, even though the weather is their main business.

Another service provided by a few FSSs (and some towers) is a *DF steer* (that stands for Direction Finding). DF does what the name implies—finds direction for lost aircraft. The pilot will usually be asked to transmit a count from 1 to 10 on the mike. The FSS has an ADF receiver that points to the aircraft and will help the FSS bring you home. It's a "lost procedure" that works without radar or other airborne equipment. All you need is a transmitter and receiver. In the Airport Directory, this airport will be designated as a VHF/DF.

While you've got your chart out, go back to the legend and note how the FSS frequencies are given. For instance, the heavy line box giving the FSS data signifies two unstated frequencies: 121.5 (emergency) and 122.2 (standard FSS) plus a couple of standard military frequencies. Any other frequencies available at the FSS must be shown above the box, such as 123.6, etc. Note the 122.1R that appears over most VOR boxes—this means that FSS will *receive* you on 122.1 and transmit back on the VOR. In other words, most VORs serve a secondary function as a "remote outlet" for the FSS, which gets us past a lot of line-of-sight problems common to the VHF communications.

Shaded upper corners in the box indicate En Route Flight Advisory Service, or EFAs, which is called "Flight Watch" in pilot talk. Call Flight Watch on 122.0 for information regarding actual weather and thunderstorm activity along your route anytime between 6 A.M. and 10 P.M. A small shaded box located in the bottom right corner of the VOR box or NDB box indicates a Transcribed Weather Broadcast (TWEB) is available on the navaid frequency. FSS puts together the tape recording of current and/or forecast weather for the airport and/or area. The information on the TWEB may vary considerably!

Unicom

What happened to Unicom? If there's an FSS on the field, it goes back to soliciting fuel and doing small favors for the pilot. What if there's no Unicom? Then you would broadcast your traffic advisories on what is known as Multicom—122.9 MHz.

Airport Markings and Lighting Aids

The segmented circle gives the traffic pattern at a glance. In Fig. 3-24 you will see the "clear" arrow indicating the wind and the dark arrows indicating the designated turn from base-leg to final. A flashing amber light near the segmented circle indicates right hand traffic, confirmed by the landing strip indicators.

On and Near the Runways

The runway in Fig. 3-25 with an X on it is closed! In Fig. 3-26. you will see a wind tee, tetrahedron, and wind sock. The wind is coming right down the runway and this is how they look.

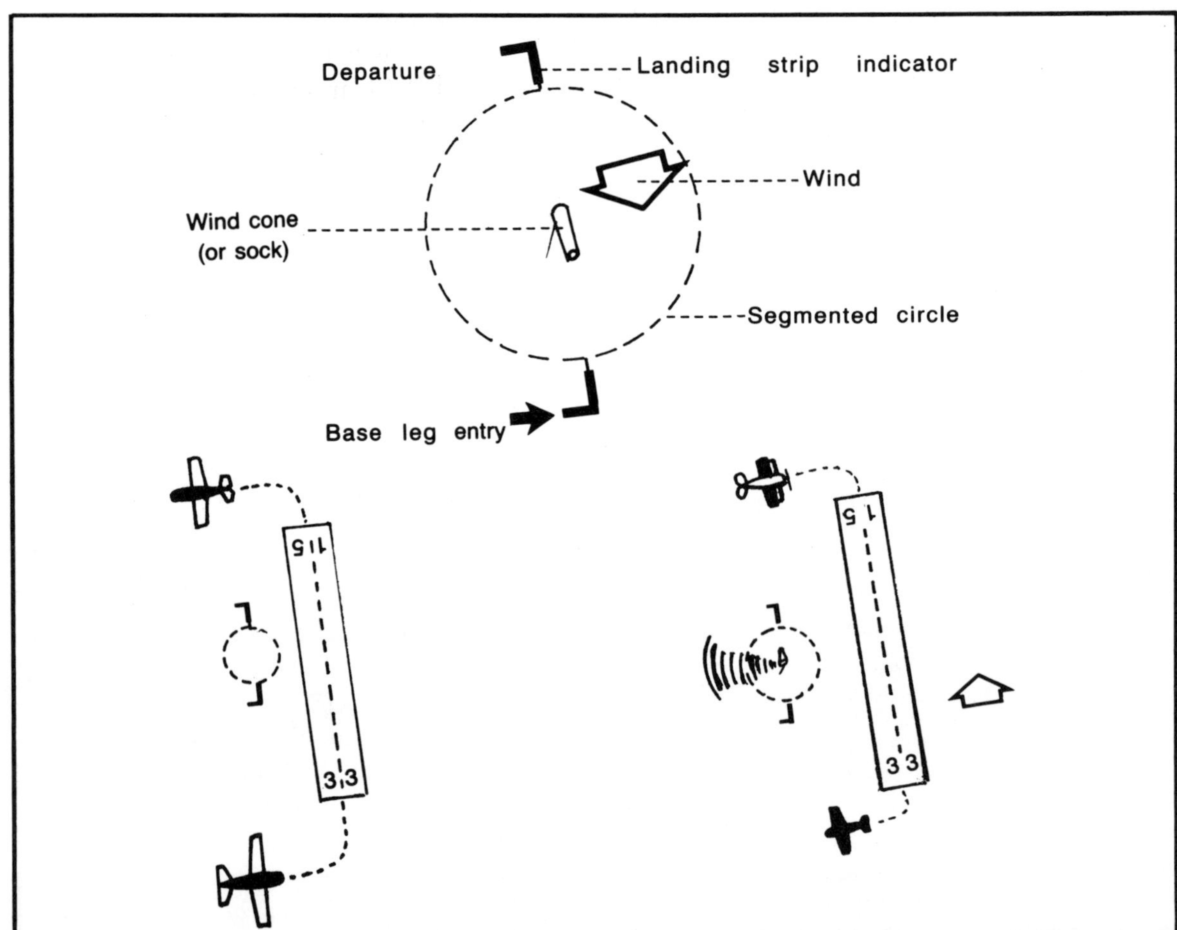

Fig. 3-24. Segmented circle.

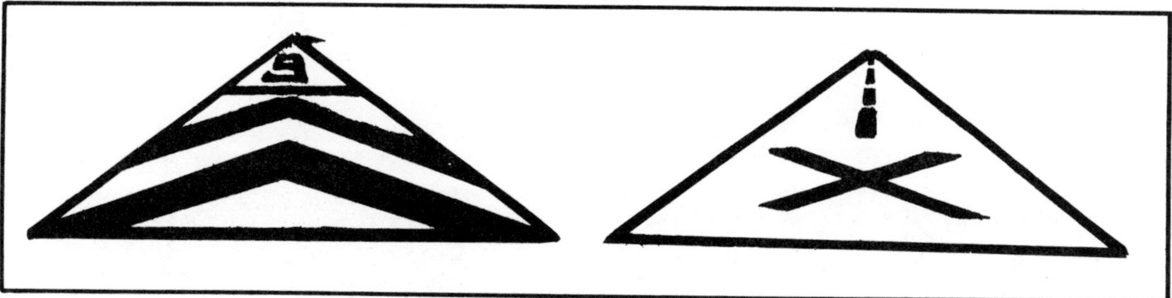

Fig. 3-25. Runway overrun, left; closed runway, right.

The big black line below the 9 (toward you) is the displaced threshold. (Just like 27 at Weld County.) A displaced threshold is usable for taxi and takeoff, but not for landing.

Notice the VASI (Visual Approach Slope Indicators) on either side of the runways in Fig. 3-27. Here's what they mean:

A. Red over white—you're all right.
B. Is impossible; forget it!
C. White over white, you're too high.
D. Red over red, you're dead (too low).

Another variety of VASI is the tricolor type. It's a single light unit that shows you a red light for too low, amber light for too high, and a green light for right on glideslope.

While we're on the subject of lights, Fig. 3-28 shows important difference in the rotating beacons found at most airports.

If an airport's rotating beacon is turned on during the daytime and *if the airport is in a control zone* it means the ground visibility is less than 3 statute miles and/or the ceiling is less than 1,000′ AGL. In other words, the qualified weather observer serving the control zone has made an official declaration that the entire control zone is now below VFR minimums and a Special VFR Clearance will be necessary to take off or land (see FAR 91.107). In contrast, the rotating beacon at Weld County (no control zone) is operated by a photo-cell—each pilot-in-command makes his own judgement of the weather.

Special Use Airspace

Prohibited Area: "Prohibited Areas contain airspace of defined dimensions identified by an area on the surface of the earth within which the flight of aircraft is prohibited. Such areas are established for security or other reasons associated with the national welfare. These

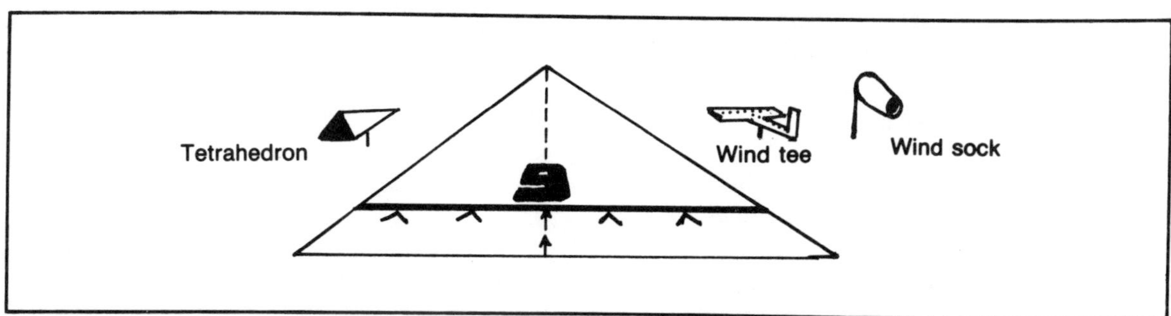

Fig. 3-26. Displaced threshold.

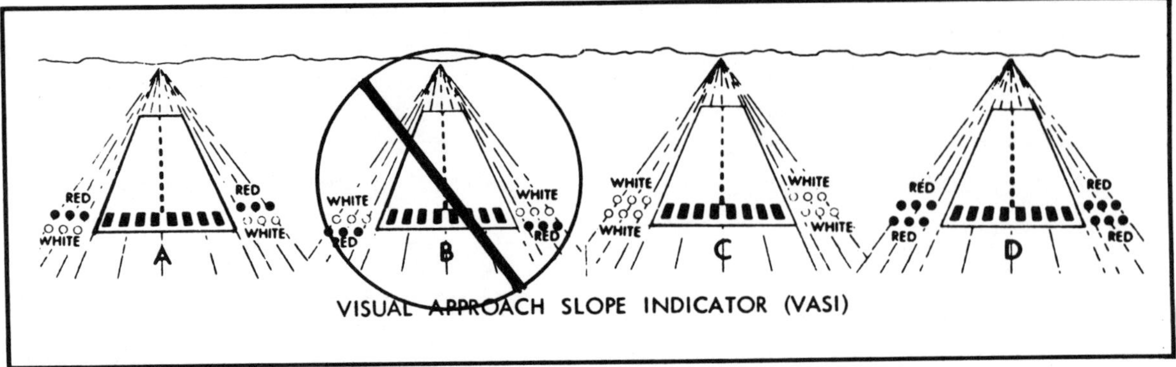

Fig. 3-27. VASI lights.

areas are published in the Federal Register and are depicted on aeronautical charts." (*AIM* para. 111) The White House and Pentagon are Prohibited Areas.

Restricted Area: "Restricted Areas contain airspace identified by an area on the surface of the earth within which the flight of aircraft, while not wholly prohibited, is subject to restrictions. Activities within these areas must be confined because of their nature or limitations imposed upon aircraft operations that are not a part of those activities or both. Restricted areas denote the existence of unusual, often invisible, hazards to aircraft such as artillery firing, aerial gunnery, or guided missiles. Penetration of Restricted Areas without authorization from the using or controlling agency may be extremely hazardous to the aircraft and its occupants." (*AIM* para. 112)

Warning Areas are located over international waters and contain unusual, often invisible hazards such as aerial gunnery or guided missiles.

Military Operations Areas (MOA):

- ☐ a. MOAs consist of airspace of defined vertical and lateral limits established for the purpose of separating certain military training activities from air traffic.
- ☐ b. Some training activities may necessitate acrobatic maneuvers, and the United States Air Force (USAF) is excempted from the regulation pro-

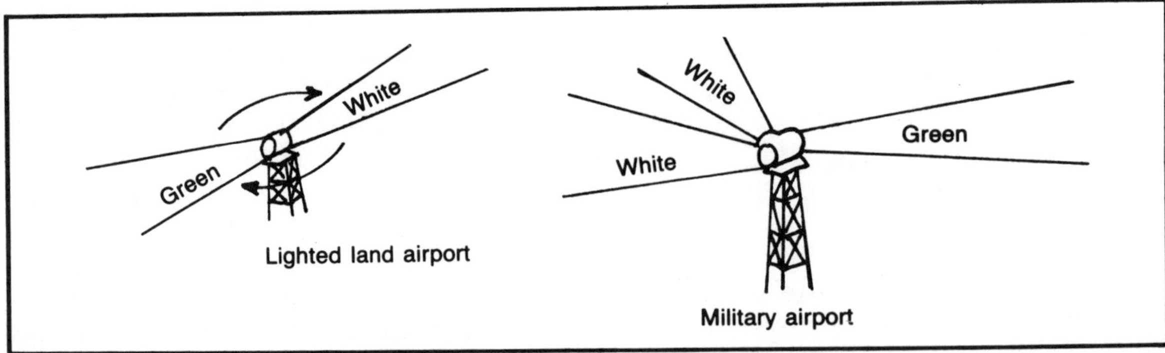

Fig. 3-28. Civil (left) and military (right) beacons.

hibiting acrobatic flight on airways within MOAs.

☐ c. Pilots operating under VFR should exercise extreme caution while flying within a MOA when military activity is being conducted. Information regarding activity in MOAs may be obtained from any FSS within 100 miles of the area.

☐ d. These areas are depicted on Sectional Charts. (*AIM* para. 114)

Military Training Routes:

☐ a. . . . To be proficient, the military services must train in a wide range of airborne tactics. One phase of this training involves "low level" combat tactics. The required maneuvers and high speeds are such that they may occasionally make the see-and-avoid aspect of VFR flight more difficult without increased vigilance in areas containing such operations . . .

☐ b. . . .MTR routes are mutually developed for use by the military for the purpose of conducting low-altitude, high-speed training . . .

☐ c. . . . Generally, MTRs are established below 10,000 feet MSL for operations at speeds in excess of 250 knots . . .

☐ d. Military Training Routes will be identified and charted as follows:

—(a) (MTR routes) at or below 1,500 feet AGL . . . will be identified by four digit numbers; e.g. IR 1006, VR 1007, etc.

—(b) IR and VR above 1,500 feet AGL (segments of these routes may be below 1,500) will be identified by three digit numbers; e.g. IR 008, VR 009, etc.

☐ e.Nonparticipating aircraft are not prohibited from flying within an MTR; however, extreme vigilance should be exercised when conducting flight through or near these routes. Pilots should contact FSSs within 100 nm of a particular MTR to obtain current information or route usage in the vicinity . . . Route width varies each MTR and can extend several miles on either side of the charted MTR centerline. (Excerpted from *AIM* para. 132)

CROSS-COUNTRY FARS

Here are some important regulations:

FAR 91.5 Preflight Action

Each pilot in command shall, before beginning a flight, familiarize himself with all available information concerning that flight. This information must include:

(a) For a flight under IFR or a flight not in the vicinity of an airport, weather reports and forecasts, fuel requirements, alternatives available if the planned flight cannot be completed, and any known traffic delays of which he has been advised by ATC.

(b) For any flight, runway lengths at airports of intended use, and the following takeoff and landing distance information:

(1) For civil aircraft for which an approved airplane flight manual containing takeoff and landing distance data is required, the takeoff and landing distance data contained therein; and

(2) For civil aircraft other than those specified in subparagraph (1) of this paragraph, other reliable information appropriate to the aircraft, relating to aircraft performance under expected values of airport elevation and runway slope, air-

craft gross weight, and wind and temperature.

FAR 91.22 Fuel Requirements

(a) No person may begin a flight in the airplane under VFR unless (considering wind and forecast weather conditions) there is enough fuel to fly to the first point of intended landing and, assuming normal cruising speed—

(1) During the day, to fly after that for at least 30 minutes; or

(2) At night, to fly after that for at least 45 minutes.

FAR 91.24 ATC Transponder Equipment

(b) 1. Group I TCA (and Mode C); 2. Group II TCA; 3. Group III TCA; and 4 . . . In all controlled airspace of the 48 contiguous states and D.C. above 12,500 feet MSL excluding airspace at and below 2500′ AGL (and Mode C).

FAR 91.65 Operating near Other Aircraft

(a) No person may operate an aircraft so close to another aircraft as to create a collision hazard.

(b) No person may operate an aircraft in formation flight except by arrangement with the pilot in command of each aircraft in the formation.

(c) No person may operate an aircraft carrying passengers for hire, in formation flight.

(d) Unless otherwise authorized by ATC, no person operating an aircraft may operate his aircraft in accordance with any clearance or instruction that has been issued to the pilot of another aircraft for radar Air Traffic Control purposes.

FAR 91.75 Compliance with ATC Clearances and Instructions

(a) When an ATC clearance has been obtained, no pilot in command may deviate from that clearance, except in an emergency, unless he obtains an amended clearance. If a pilot is uncertain of the meaning of an ATC clearance, he shall immediately request clarification from the ATC.

(b) Except in an emergency, no person may, in an area in which air traffic control is exercised, operate an aircraft contrary to an ATC instruction.

(c) Each pilot in command who deviates, in an emergency, from an ATC clearance or instruction shall notify ATC of that deviation as soon as possible.

(d) Each pilot in command who (though not deviating from a rule of this subpart) is given priority by ATC in an emergency, shall, if requested by ATC, submit a detailed report of that emergency within 48 hours to the chief of that ATC facility.

FAR 91.77 ATC Light Signals

ATC light signals have the meaning shown in the following table [Fig. 3-29].

FAR 91.79 Minimum Safe Altitudes; General

Except when necessary for takeoff or landing, no person may operate an aircraft below the following altitudes:

(a) ANYWHERE. An altitude allowing, if a power unit fails, an emergency landing without undue hazard to persons or property on the surface.

(b) OVER CONGESTED AREAS. Over any congested area of a city, town, or settlement, or over any open air assembly of persons, an altitude of 1,000 feet above the highest obstacle within a horizontal radius of 2,000 feet of the aircraft/.

(c) OVER OTHER THAN CONGESTED AREAS. An altitude of 500 feet above the sur-

Color and type of signal	Meaning with respect to aircraft on the surface	Meaning with respect to aircraft in flight
Steady green......	Cleared for takeoff....	Cleared to land.
Flashing green.....	Cleared to taxi.......	Return for landing (to be followed by steady green at proper time).
Steady red.........	Stop................	Give way to other aircraft and continue circling.
Flashing red.......	Taxi clear of runway in use	Airport unsafe—do not land.
Flashing white......	Return to starting point on airport.	Not applicable.
Alternating red and green.	Exercise extreme caution.	Exercise extreme caution.

Fig. 3-29. ATC light signals.

face, except over open water or sparsely populated areas. In that case, the aircraft may not be operated closer than 500 feet to any person, vessel, vehicle, or structure.

FAR 91.87 Operation at Airports with Operating Control Towers.

(a) GENERAL. Unless otherwise authorized or required by ATC, each person operating an aircraft to, from, or on an airport with an operating control tower shall comply with the applicable provisions of this section.

(b) COMMUNICATIONS WITH CONTROL TOWERS. No person may, within an airport traffic area, operate an aircraft to, from, or on an airport having a control tower . . . unless two-way radio communications are maintained between that aircraft and the control tower. However, if the aircraft radio fails in flight, he may operate that aircraft and land if weather conditions are at or above basic VFR weather minimums, he maintains visual contact with the tower, and he receives a clearance to land.

(c) CLEARANCES REQUIRED. No person may, at an airport with an operating control tower, operate an aircraft on a runway or taxiway, or take off or land an aircraft, unless an appropriate clearance is received from ATC. A clearance to "taxi to" the takeoff runway assigned to the aircraft is not a clearance to cross that assigned runway at any point, but is a clearance to cross other runways that intersect the taxi route to that assigned takeoff runway. A clearance to "taxi to" any point other than an assigned takeoff runway is a clearance to cross all runways that intersect the taxi route to that point.

FAR 91.89 Operation at Airports without Control Towers

Each person operating an aircraft to or from an airport without an operating control tower shall—

(a) In the case of an airplane approaching to land, make all turns of that airplane to the left unless the airport displays approved light signals or visual markings indicating that turns should be made to the right, in which case the pilot shall make all turns to the right;

(b) In the case of a helicopter approaching to land, avoid the flow of fixed-wing aircraft; and

(c) In the case of an aircraft departing the airport, comply with any FAA traffic pattern for that airport.

FAR 91.90 Terminal Control Areas

(a) GROUP I TERMINAL CONTROL AREAS

(1) OPERATING RULES. No person may operate an aircraft within a Group I terminal control area . . . . except in compliance with the following rules:

(i) No person may operate an aircraft within a Group I terminal control area unless he has received an appropriate authorization from ATC prior to the operation of that aircraft in that area.

(2) PILOT REQUIREMENTS. The pilot in command of a civil aircraft may not land or take off that aircraft from an airport within a Group I terminal control area unless he holds at least a private pilot certificate.

(3) EQUIPMENT REQUIREMENTS. Unless otherwise authorized by ATC . . . . no person may operate an aircraft within a Group I terminal control area unless that aircraft is equipped with—

(i) An operable VOR . . . receiver,

(ii) An operable two-way radio capable of communicating with ATC on appropriate frequencies for that terminal control area; and

(iii) An operable Mode C transponder with encoding altimeter)

(b) GROUP II TERMINAL CONTROL AREAS.

(1) OPERATING RULES. No person may operate an aircraft within a Group II terminal control area . . . . . except in compliance with the following rules:

(i) No person may operate an aircraft within a Group II Terminal Control ARea unless he has received an appropriate authorization from ATC prior to operation of that aircraft in that area, and unless two-way radio communications are maintained, within that area, between that aircraft and the ATC facility.

(2) EQUIPMENT REQUIREMENTS. Unless otherwise authorized by ATC . . . . . . no person may operate an aircraft within a Group II terminal control area unless that aircraft is equipped with—

(i) An operable VOR . . . receiver.

(ii) An operable two-way radio capable of communicating with ATC on the appropriate frequencies for that terminal control area; and

(iii) (An operable Mode A transponder.)

FAR 91.105 Basic VFR Weather Minimums

(a) except as provided in 91.107, no person may operate an aircraft under VFR when the flight visibility is less, or at a distance from clouds that is less, than that prescribed for the corresponding altitude in the following table [Fig.3-30].

(b) Except as provided in 91.107, no per-

Altitude	Flight visibility	Distance from clouds
1,200 feet or less above the surface (regardless of MSL altitude)—		
Within controlled airspace..........	3 statute miles.....	500 feet below. 1,000 feet above. 2,000 feet horizontal.
Outside controlled airspace.........	1 statute mile as provided in § 91.105(b).	Clear of clouds.
More than 1,200 feet above the surface but less than 10,00 feet MSL—		
Within controlled airspace..........	3 statute miles.....	500 feet below. 1,000 feet above. 2,000 feet horizontal.
Outside controlled airspace.........	1 statute mile......	500 feet below. 1,000 feet above 2,000 feet horizontal.
More than 1,200 feet above the surface and at or above 10,000 feet MSL.	5 statute miles.....	1,000 feet below. 1,000 feet above. 1 mile horizontal.

Fig. 3-30. Visibility and cloud clearance minimums.

son may operate an aircraft, under VFR, within a control zone beneath the ceiling when the ceiling is less than 1,000 feet.

(c)Except as provided in 91.107, no person may take off or land an aircraft, or enter the traffic pattern of an airport, under VFR, within a control zone—

(1) Unless ground visibility at that airport is at least three statute miles; or

(2) If ground visibility is not reported at that airport, unless flight visibility during landing or take off, or while operating in the traffic pattern, is at least three statute miles.

(e) For the purposes of this section, an aircraft operating at the base altitude of a transition area or control area is considered to be within the airspace directly below that area

FAR 91.107 Special VFR Weather Minimums

(a) . . . when a person has received an appropriate ATC clearance, the special weather minimums of this section instead of those contained in 91.105 apply to the operation of an aircraft by that person in a control zone under VFR.

(b) No person may operate an aircraft in a control zone under VFR except clear of clouds.

(c) No person may operate an aircraft in a control zone under VFR unless flight visibility is at least one statute mile.

(d) No person may take off or land an aircraft at any airport in a control zone under VFR—

(1) Unless ground visibility at that airport

is at least one statute mile: or

(2) If ground visibility is not reported at that airport, unless flight visibility during landing or takeoff is at least one statute mile.

(e) No person may operate an aircraft in a control zone under the special weather minimums of this section, between sunset and sunrise (unless IFR rated and equipped).

FAR 91.109 VFR Cruising Altitude or Flight Level

. . . each person operating an aircraft under VFR in level cruising flight more than 3,000 feet above the surface shall maintain the appropriate altitude or flight level prescribed below, unless otherwise authorized by ATC:

(a) When operating below 18,000 feet MSL and —

(1) On a magnetic course of zero degrees through 179 degrees, any odd thousand foot MSL altitude + 500 feet (such as 3,500, 5,500, or 7,500); or

(2) On a magnetic course of 180 degrees through 359 degrees, any even thousand foot MSL altitude +500 feet (such as 4,500, 6,500, or 8,500).

Chapter 4

Meteorology

Weather is the state of the atmosphere at any given time. Since flight takes place in the air, weather and flight are so closely related that they are inseparable. It is therefore the responsibility of the pilot to learn as much about the weather as possible. The pilot must fully understand that he is the final word on whether flight should be continued. If this involves a judgment based on weather, then that judgment should be based on knowledge and understanding.

Some pilots don't realize it, but *they* are the "weatherman," not the person on the ground. Flight involves a series of continued judgments and evaluations about the situation at hand. The pilot must look out of the cockpit, observe, evaluate, and react in a manner that is safe to all concerned. That enormous responsibility weighs heavily on the shoulders of the pilot and it is a sobering thought when you consider the consequences of a poor decision. You might think that weather is a life and death situation 60 seconds out of every minute you fly! When you get down to the bottom line, *you* are the forecaster—simply because you are the one who looks out of the cockpit, observes, and makes the continuing decision "go or no-go."

A NEW KIND OF TIME

There are only two kinds of pilots—survivors and the other kind. To ensure yourself a long and happy career in aviation, you must learn to approach the weather both as your best friend and as your deadliest enemy. We all come to know and love our friends, but a survivor also knows his enemy—and knows him well. This will be your introduction to a lifelong study.

Now picture an airline pilot checking the weather for a trip from New York to Tokyo and being told that "the forecast for Japan is thunderstorms at noon, lasting until dusk."

Such times as these are useless to someone halfway around the world, or even a few time zones away. Those who live and work in one time zone need a time when "noon" is about the middle of the day, but a pilot needs to eliminate time zones altogether so that there is only *one* time zone all over the globe. This has been done since the days of sailing ships by simply accepting the time in Greenwich, England (a suburb of London) as being the universal and worldwide time. We call it Greenwich Mean Time (GMT) or Zulu time. Greenwich is also the location of the zero line of longitude, and roughly every 15 degrees from zero is another time one (360 degrees of longitude divided by 15 equals 24—the number of hours in each day).

In Fig. 4-1 you can see that London and Denver are separated by seven hours (time zones). Since the earth rotates to the east, we are seven hours *behind* London, so that when it's 5 A.M. here, the British are going on lunch break (5 A.M. + 7 hours = 12 noon Zulu time). To make the math easier, we use the military time clock of 24 hours—noon is spoken as "twelve hundred hours," then comes "thirteen hundred" (1 P.M.), "fourteen hundred," etc., right on up to midnight, which is "twenty-four hundred hours." Think of it the easy way:

Colorado time		Conversion to Zulu		Zulu time
0500	+	0700	=	1200

Therefore, to get Zulu in the Mountain Standard Time Zone, simply add seven hours. Omaha (Central Time Zone) adds six hours because they're closer to London. In the summer, when we go to Daylight Saving Time, we will be adding the six hours for Mountain Daylight Time, and Omaha will switch to five hours for Central Daylight Time. Now, if Zulu is given, how do you convert back? Just reverse the process and subtract:

Zulu Time		Conversion to MDT		Colorado time
2200	−	0600	=	1600 (4 P.M.)

That will take some practice, but you can get it—try setting your watch to Zulu time for a few days.

THE SKY AND CLOUDS

Figure 4-2 shows some of the various types of clouds. Clouds can be classified by the way they form. If the air is relatively *unstable* (caused by warming from below or high mois-

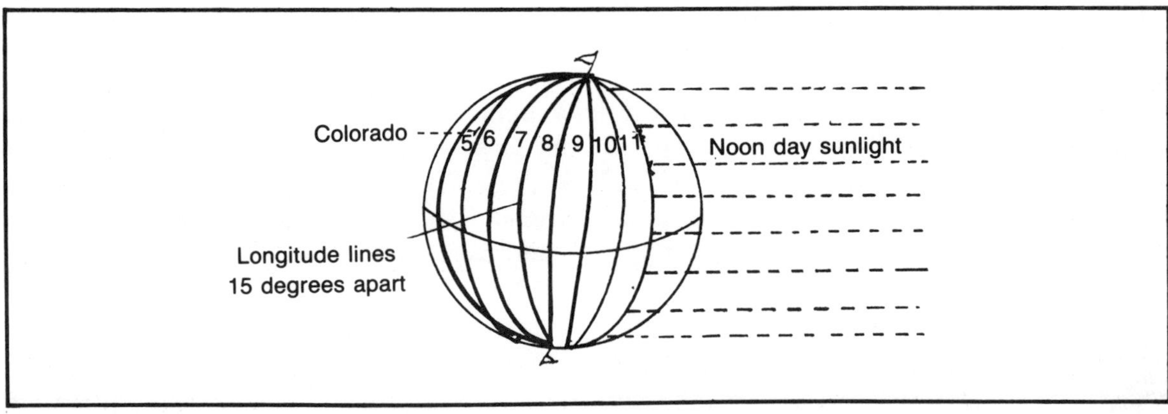

Fig. 4-1. Time zones.

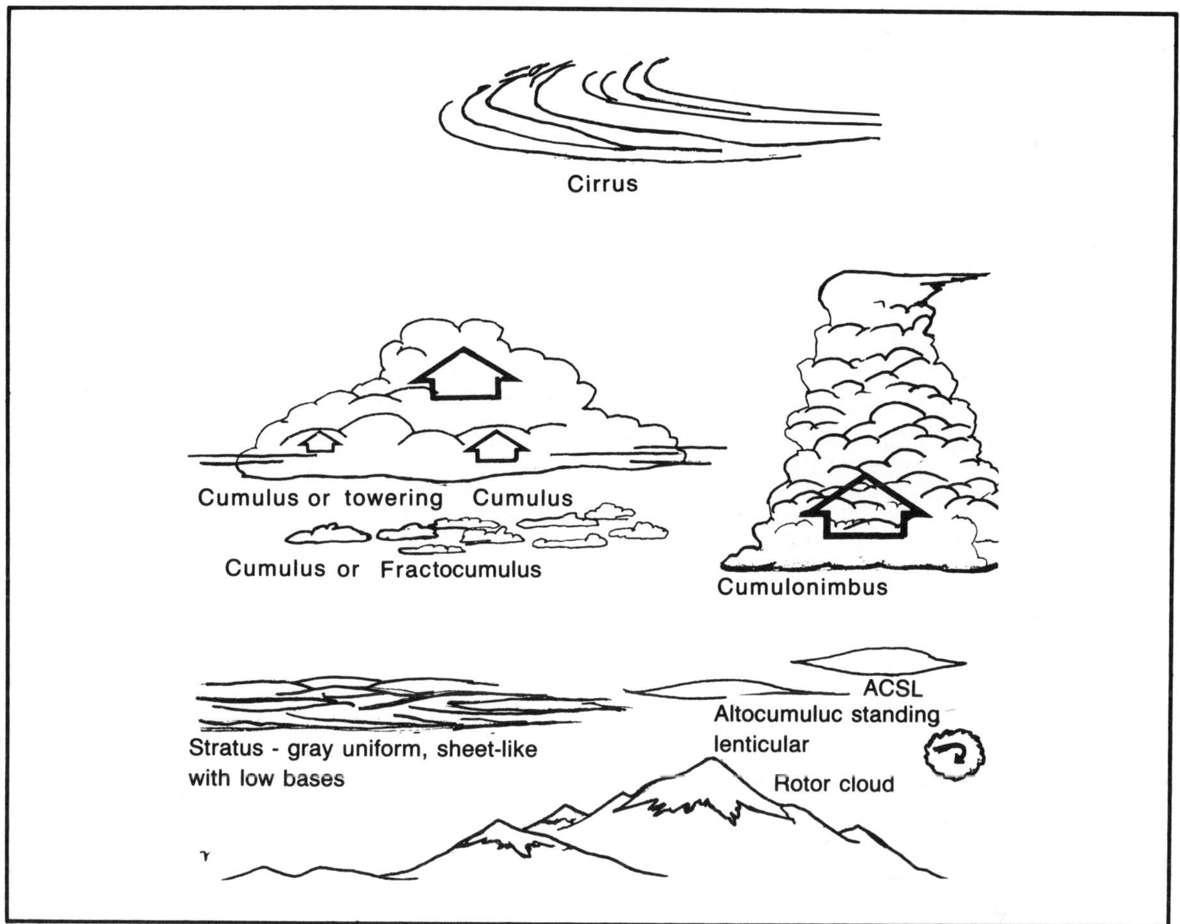

Fig. 4-2. Cloud types.

ture content), it will tend to rise. This rise in altitude will cause a temperature decrease and whatever moisture is present will condense into tiny but visible droplets.

The temperature at which this *saturation* occurs is known as the *dewpoint*, and visible moisture will form when the temperature and dewpoint get within about four degrees (or less) of each other. On the ground it's fog or low clouds, but if it forms in the sky from rising air, it will be a cloud that is flat on the bottom and puffy or bumpy on top. We call that particular shape a *cumulus* cloud.

Especially unstable air forms *towering cumulus* and *cumulonimbus*, where the updrafts become more severe and, in the case of the cumulonimbus, or thunderstorm, the cloud turns dark because of the massive quantities of water droplets it contains. In summertime, these are commonly due to *convective turbulence*. You have seen wisps of rain falling from clouds that evaporates before it hits the surface. This is called *virga* and it is common to cumulus clouds.

Clouds formed in *stable* air have a flatter, more layered appearance that we call *stratiform* or *stratus*. High water content in these create *nimbostratus* (the *nimbo* means rain). Clouds at a medium height are prefixed *alto*. Stability of the air is determined by the rate at which the air cools with altitude, called *lapse rate*.

Cirrus clouds are formed in the very cold air at high altitudes. They are composed of ice crystals and have a thin, wispy appearance giving them the nickname "horsetails." We can now divide clouds into their four "families" according to their height:

- ☐ A *cirro* prefix means high.
- ☐ *Alto* means medium altitude.
- ☐ No prefix, such as *stratus* or *cumulus*, means low altitude.
- ☐ Clouds with extensive vertical development, such as *cumulonimbus*.

An almond or lens-shaped cloud is called a *standing lenticular* and is unique to mountainous areas (Fig. 4-). This "standing" cloud will not drift with the wind but is actually stationary, because it is forming above the mountain peaks (where rising air brings the tempo/dewpoint together) and dissipating well downwind of the ridgeline (where the descending air warms so that the temp/dewpoint spread apart and the water evaporates back into its vapor form). That rising and falling of the air is caused by strong winds aloft (50 knots or more) stretching the lenticulars out downwind (leeward) as the air flows up, over, and down the mountain slopes. The air near these clouds contains violent downdrafts, so stay away from such areas until you're an accomplished "mountain pilot." When crossing a ridge line at low altitude, the greatest potential danger—caused by descending air currents—will usually be encountered on the leeward side when flying into the wind (toward rising terrain).

We will use the cloud types to indicate what kind of weather is in the vicinity (or upwind), but let's first see how the Flight Service Station will measure the altitude and quantity of clouds for us. Remember that FAR 91.105 prohibits the VFR pilot from flying in or too near the clouds. The sky appearance will be classified as follows:

☐ Clear (CLR)	Less than .1 sky coverage by clouds.
☐ Scattered (SCT)	.1 to .5 (half) sky coverage.
☐ Broken (BKN)	.6 to .9 sky coverage.
☐ Overcast (OVC)	More than .9 coverage.
☐ Obscured (X)	A surface-based obstruction to venting a sky vision such as fog or smoke, is preventing a skyview.

Do you remember that basic VFR traffic is prohibited from operating in a Control Zone when the ceiling is less than 1,000 feet? Well, a ceiling is the lowest layer of clouds that is reported as broken, overcast, or obscured and is not classified as "thin" or transparent. If you can see through a solid overcast, it is not a legal ceiling.

Ceilings are always measured in hundreds of feet Above Ground Level (AGL) and this is important knowledge to the pilot, since other clouds are often forecast above Mean Sea Level. If Denver predicts stratus at 6,000, you'd better know which way they measured it!

BAROMETRIC PRESSURE

A major importance of atmospheric pressure is in the measurement of day-to-day changes in the pressure (primarily due to unequal heating of the Earth's surface) and movement of the air masses that give us weather.

Watch the weather map on television (preferably the pilot's weather program called *A.M. Weather* on the local PBS station) and you will pick up the regular flow and movement of regions of high and low pressure that are variations above and below the 29.92 inch standard (at sea level) pressure.

A *high* is a large mass of higher than standard pressure air, which you can visualize as a huge dome or mountain of thicker air pressing down against the surface of the Earth, where the air tends to flow away from the center in an outward direction, but rotating in a clockwise direction. High pressure areas are usually good flying weather, although smaller scale problems, such as fog or thunderstorms, can still occur inside the high. But the air is ordinarily stable, tending to resist upward movements or even sinking.

Here is another part of the VFR weather minimums; remember 3 miles for controlled airspace, 1 mile for uncontrolled, and 5 miles when above 10,000 MSL? The visibility will always be measured in statute miles (and sometimes fractions thereof). Whether reported or forecast, we should know what is restricting the visibility, and there are several possibilities:

- ☐ D Dust.
- ☐ S Snow.
- ☐ SW Snow showers.
- ☐ ZR Freezing rain.
- ☐ BS Blowing snow.
- ☐ F Fog.
- ☐ R Rain.
- ☐ RW Rain showers.
- ☐ T Thunderstorm.
- ☐ H Haze.
- ☐ L Drizzle.
- ☐ K Smoke.
- ☐ A Hail.

The abbreviations for sky condition and restrictions to visibility will be important for you to know when we start reading the weather reports and forecasts.

Low pressure areas are regions where the air is thin and unstable, tending to rise in vertical currents that may cause more violent forms of weather. A falling barometric pressure has always been taken as a danger sign, primarily because it may signal the approach of a *front*. Lows seem to form in the shape of a valley or whirlpool (although remember that the air is rising), and are rotating counterclockwise, drawing air in toward the center.

These differences in pressure are what creates wind. The air naturally moves from a high to a low pressure area but is diverted from a straight line flow because of the rotation of the Earth (*Coriolis Effect*) into the characteristic clockwise and counterclockwise direction about the highs and lows. A pilot must remember that wind velocity is always measured in knots, and that wind direction is always *from* (never to) but can be expressed in either True or Magnetic directions. Your best rule of thumb is that weather reports and forecasts in their written or "paper" form will have the wind in True directions (meteorologists don't work in Magnetic), but a wind broadcast directly to the pilot, such as an Airport Advisory from a Control Tower or Flight Service Station, will be converted to Magnetic direction to match up with the runway.

TEMPERATURE, DEWPOINT, AND FOG

Remember that when the temperature and dewpoint come within four degrees or less of each other, any moisture present in the air will be near the saturation point and will likely condense to a visible form, such as rain, snow, or fog. The form of visibility problem we get depends on the actual temperature at which the temp/dewpoint are nearly equal. But let's examine fog more closely. This restriction to visibility usually occurs by one of the following three cooling processes.

Upslope fog is caused by air that cools as a breeze blows up a mountain slope, or any rising terrain. Since the air will cool steadily, it may eventually meet the dewpoint and cause visible moisture. If this happens at the surface, we call it upslope fog. Of course, an observer looking up at the mountain would probably say there were "clouds topping the ridges," but to the person standing on the peak, that is fog! If the visible moisture occurs above the surface, it'll form a cloud—stratus if the air is stable, cumulus if it is unstable.

Advection fog results from moist air condensing as it moves over a cooler surface. Picture a coastline with a gentle wind coming in off the water. That air is well loaded with moisture, especially if it's fairly warm (high dewpoint). If the ground along the coastal region is cooler than the water, then the air will cool as it blows inland, and the high dewpoint with the lowering temperature gives the usual result. A stronger wind (15 mph or better) will eliminate the fog by lifting it aloft. Think of coastal fog in Seattle or Chicago or wherever you've seen it, but actually, advection fog can extend great distances inland, as in the case of midwestern fogs off the Gulf coast.

Radiation fog is the tough one to understand, but perhaps the most common. Picture a clear warm day with the Earth soaking up a maximum of heat from the sun. After nightfall, this process reverses and the Earth begins radiating away all of that heat (winds are calm or light). By about dawn, which is the coldest part of the day, all of the warmth has lifted up a short distance (forming a *temperature inversion*) and left the belt of air right at the ground at its coldest temperature. If there is sufficient moisture present to allow that temperature to reach the dewpoint, a fog or low overcast occurs, especially in the low valleys. Its common name is *ground fog* or *morning fog*.

STRUCTURAL ICE

Here's another problem that the noninstrument rated pilot should avoid at all times. Any form of ice or frost accumulation on the aircraft can increase your weight and drag, while decreasing your thrust and lift—the aerodynamic shapes of the wing and propeller are critical to the aircraft's performance.

Frost is an ice crystal that forms when the temperature of the collecting surface is at or below the dewpoint of the adjacent air and the dewpoint is below freezing. Frost that formed on the aircraft overnight should, at the very least, be scrubbed smooth before flying. In sufficient accumulations it will cause early separation of the airflow over a wing surface and prevent the airplane from becoming airborne.

Structural icing is best avoided by simply staying out of visible moisture of any kind when the outside air is near or below freezing temperatures. Freezing rain is most likely to have the highest accumulation rate.

Structural ice takes two different forms (although they are often mixed):

Rime ice, like refrigerator ice, is milky in appearance and granular in texture. This is formed from the small water droplets characteristics of stratus clouds or fog as they hit the aircraft and freeze instantly, trapping air between the ice "pellets" and giving rime its distinctive appearance.

Clear ice is a glossy glaze that forms from larger water droplets (as in cumulus) that freeze more slowly as they flow back over your wings, making a solid mass of ice that tends to conform to the airfoil shape. The most rapid accumulations are usually encountered at temperatures from 0 to – 15 degrees Celsius.

THUNDERSTORMS

It goes without saying that a thunderstorm (cumulonimbus or Cb) is very dangerous to light aircraft. Exceeding the airplane's structural strength is just one of the major hazards found in Cbs, and this simply means that the violent turbulence inside has been known to break up aircraft.

It is important to understand the three stages of a Cb's life (Fig. 4-3). The first, *cumulus*, is the building stage and features updrafts that can extend from near the surface to the top of the cloud. Another way of thinking of it is that every cumulus cloud is trying to form into a thunderstorm. The presence of lightning indicates that it has succeeded. Fortunately, only a few Cus grow up to be Cbs, but a sky of cumulus always indicates a bumpy ride.

The second stage, *mature* begins as the first rain reaches the Earth's surface. This means that, while the updrafts are continuing and even accelerating, the water droplets in the cloud have grown to a size too large for the updrafts to hold, so gravity pulls these

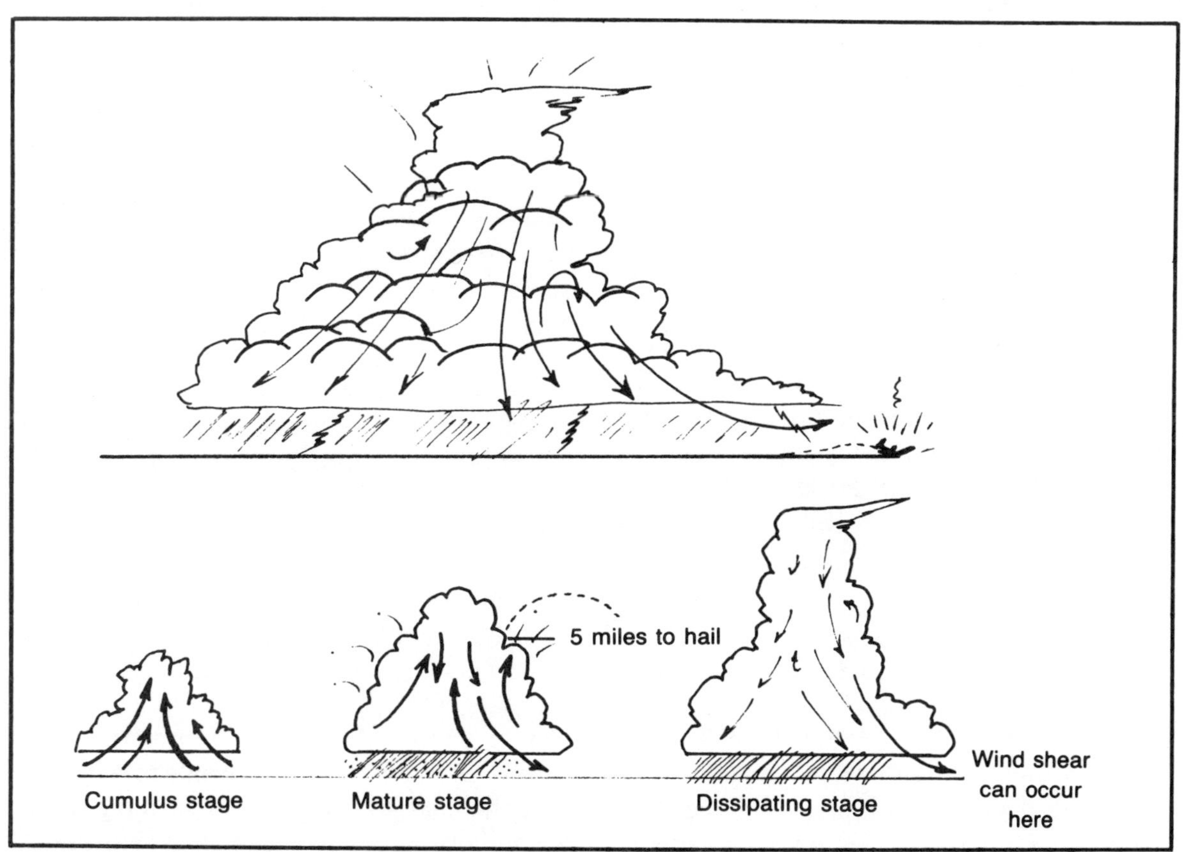

Fig. 4-3. The three stages of a thunderstorm.

larger droplets out and this drags the air with it. This is how the downdrafts begin. The up and down drafts existing together create the *vertical wind shear*, which can give severe to extreme turbulence.

The second major hazard of thunderstorms, *hail*, is now forming as ever-growing hailstones ride the up/down drafts back and forth across the freezing level. Large hail results from strong updrafts and large water content. The hail tends to work its way across the storm with the wind, and has been thrown out of the storm on the downwind side up to five miles. I shouldn't have to tell you what hail can do to aluminum leading edges of the aircraft that attempts to fly on the downwind side of the storm or in the storm itself. Tops of the storm may reach to 60,000 feet or higher.

As the Cb progresses into its *dissipating* stage, it will show the "anvil" shape that forms at the top of the Cb and streaming downwind, showing you the direction of the hail as well as the direction that the storm is moving. This final stage is also characterized by complete downdrafts as the storm dumps its accumulation of water and hail. When the downdrafts hit the ground and plow outward from the storm, they present a new hazard of wind shear to the unwary pilot attempting to beat the storm into or out of an airport. This "gust front" can precede the actual storm by 15 nautical miles.

WIND SHEAR

The FAA defines *wind shear* as a sudden change in direction and/or velocity across a very short distance in the atmosphere. This can create a significant problem for the pilot (even aside from the turbulence present in the shear zone) if the aircraft moves from a headwind into a tailwind. The result is a temporary loss of airspeed and lift as the mass of the aircraft has to "catch up" with the new wind situation. It lasts only several seconds, but crashes have occurred when aircraft taking off (into the wind) have suddenly transitioned into a tailwind in the vicinity of a thunderstorm.

Wind shear can also occur if strong winds are blowing over the top of a low-level inversion. You can be relatively certain of a shear zone in the inversion if you know the wind at 2,000 to 4,000 feet is 25 knots or more. In any case, the pilot's best action would be to plan a takeoff (or landing) in such a direction as to ensure transitioning to a headwind instead of to a tailwind.

FRONTS AND FRONTAL MOVEMENT

Remember that in the pressure systems discussion we pointed out the counterclockwise airflow around a low pressure area and compared it to a whirlpool of unstable air. However, the low is not usually of an even circular shape, and often elongates or stretches out to form a *trough*. In most cases, a front forms along the line of the trough and also rotates counterclockwise around the low. The front is simply the zone of transition between two masses of air that have different characteristics. We must examine the changes in temperature, dewpoint, pressure, and wind that can occur right at the frontal boundary and for hundreds of miles on either side.

The temperature change with a front is probably the easiest to think about, since the new or incoming air mass will be either colder or warmer than the air mass that is replacing. We name a front by the temperature change expected, either cold or warm. It will help your understanding of fronts to remember that the cold air mass will always try to stay at the bottom, while warm air will always be more willing to rise.

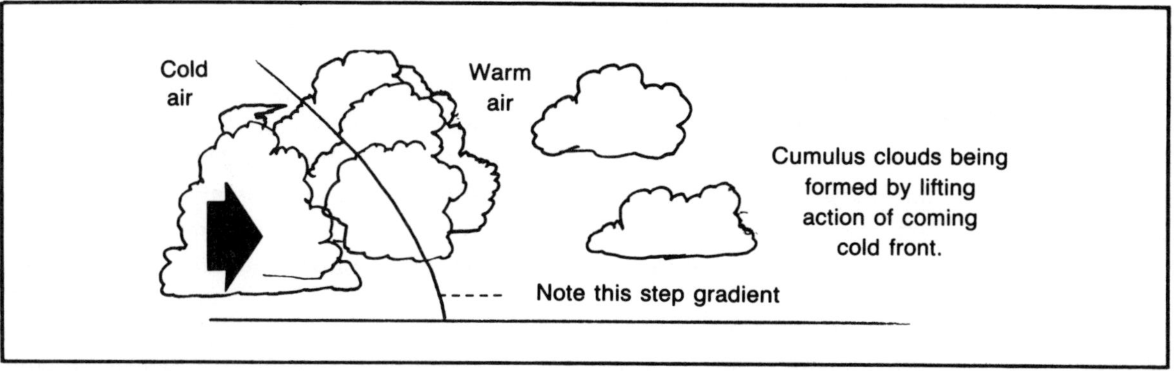

Fig. 4-4. Cold front.

The Cold Front

Here, the incoming cold air mass is thrusting under the warmer air ahead of it and pushing it aloft (Fig. 4-4). The lifting action contributes to the instability of the warmer air so that a cold front typically is marked by cumulus clouds, perhaps thunderstorms. Since cold air is more dense and has more "power," the cold front will typically move more rapidly than the warm front and clearing skies (and colder temperatures) will quickly follow. The fast-moving cold front is especially dangerous as a wave action can be set up well ahead of the front (up to 200 miles), which becomes what we call a *squall line* of the most severe thunderstorms, heavy hail, and tornados.

The Warm Front

This time, the incoming warm air mass lays across the top of the cold mass, which must still be on the bottom. So the lifting action here is more gentle than in the cold front but it is still the warm air being lifted. This means that usually the warm front is a more stable (vertically) system than the cold, but anything is possible in weather and thunderstorms *can* form in the warm front. These would be "embedded" thunderstorms, which are obscured by massive cloud layers and cannot be seen. Figure 4-5 though, shows the more typical warm front, where the clouds are all stratus types due to the lack of lifting. Since warm air is less dense, we can expect the warm front to move slower than the cold front, and they often hang around for days. This effect is exaggerated by the great width of the frontal zone (it can be over 1000 miles). Since warm air holds more water, we can expect visibility problems up to solid IFR condi-

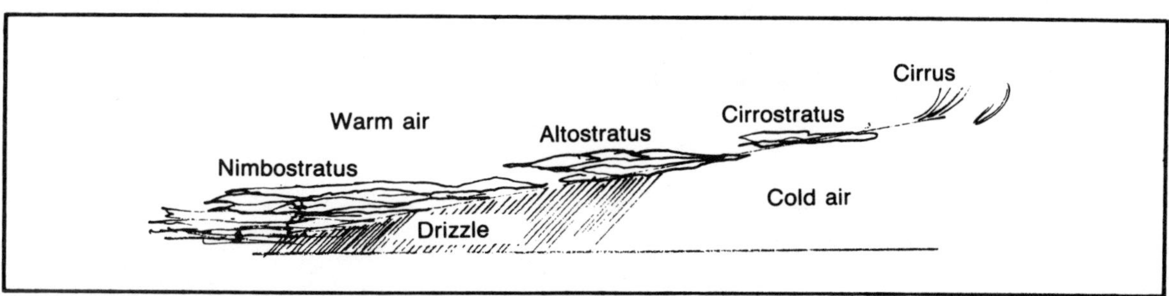

Fig. 4-5. Warm front.

tions for days, with zero ceilings where the zone reaches the ground.

Now that you've seen the cloud pictures, I've got to warn you that if both air masses are relatively dry, there will be no clouds at all in either front! But ordinarily you should concern yourself with the moisture content of the warm air mass in either front, to predict the quantity of cloud coverage.

Figure 4-6 is a look at the "satellite view" of a low with two fronts out of its center. Remember the counterclockwise rotation and notice that the somewhat circular lines show how the low has elongated into its trough form along the fronts. Those lines represent points of equal pressure connected together (similar to the terrain relief lines on a topographical map) and are called *isobars*, meaning equal barometric pressure. The trough form of the isobars along the front will show why a falling altimeter setting will signal the approach of a front, and a rising altimeter setting (or barometer) will show that the frontal line has passed.

The direction of the isobars shows us the exact direction of the winds aloft since the wind will parallel the isobars at altitude because of the Coriolis Effect. At low altitude, friction of the wind passing over the ground will slow down its velocity and interfere with the

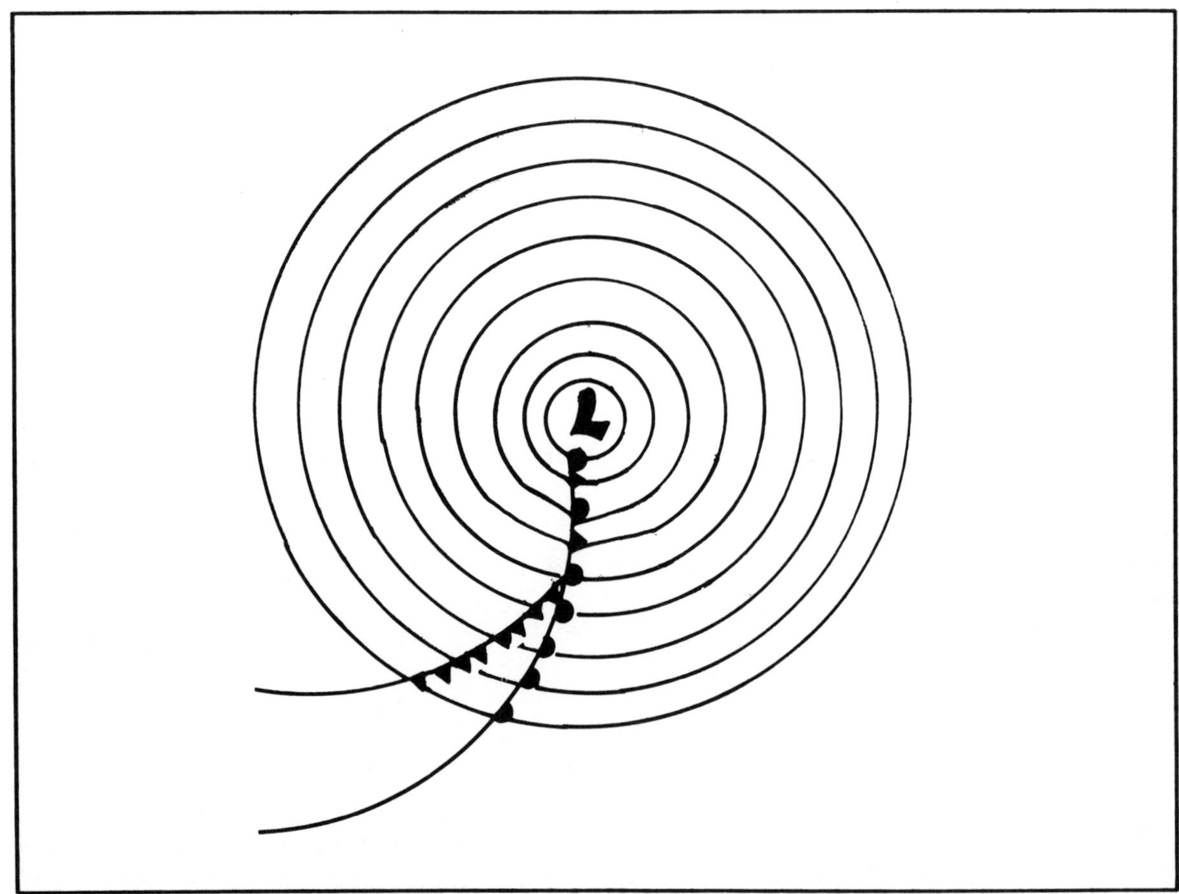

Fig. 4-6. Low pressure system and fronts.

whirlpool effect (Coriolis) enough to allow the low to draw the winds inward toward the center of the low, angling to the wind direction across the isobars, but still basically counterclockwise. The velocity of the wind can be estimated by the closeness of the isobars—where they are concentrated into a small area, that represents a rapid change of pressure, which will give a stronger flow of air. This also means that the wind always changes across a front, so we can use a shift of wind direction to indicate frontal passage.

Occluded Fronts

In Fig. 4-6, if that cold front should overtake the warm front (remember that it generally moves faster), then we will have an *occluded front*. Then all three air masses are joined together and the characteristics of both the cold front and the warm front exist at the same place.

THE REPORTING AND FORECASTING SYSTEM

Throughout the world, observers are continually recording weather data and transmitting it to certain terminal stations, where the information is used to create forecasts of future weather to be distributed back to the reporting stations for briefing pilots. This huge network of information is available to the pilot in several ways, primarily through the Flight Service Stations scattered across the country. At each FSS, an observer takes a reading of the whether every hour on the hour (and even more often if something interesting is happening) and transmits the data over a complex network to other FSSs plus the National Weather Service's (NWS) main office in Kansas City. The FSS reports, satellite transmissions, etc., to allow the meteorologists in Kansas City to predict the weather and send a forecast back to the local FSS technician so that he can inform you, the pilot, of the "best guess possible" of the future weather.

This information is in many different forms, which we will now explore. But from the outset, and through your career, you must be aware of what is a *report* of the *actual* weather being measured by a trained observer, and what is a *forecast* of conditions that *may* occur. This narrative will deal first with the reports.

The Sequence Report

The Surface Aviation (SA) or "sequence" report is a teletyped condensation of the hourly observation by the local FSS technician. Let's try one by the numbers:

DEN SA 2258 15 SCT M25 OVC 1 1/2 R-K
132/58/56/1807/993/BINOVC

It's strange looking. I know, but remember we called it a *sequence* report, and you just need to learn the order in which the information is presented. Here goes:

1. *Station identifier:* They all have 3 letters; DENver's an easy one. Not necessary to memorize, they usually make sense. The SA is for "Surface Aviation," the name of the report. An SP or RS here would indicate a "special" report—not on the hour.
2. *Observation time:* In Zulu or GMT. The 2258 tells you that this is the 2300Z observation. What time is that in Denver?
3. *Sky condition and ceiling:* We always measure cloud heights rounded off to the nearest hundred feet. You add two zeros. Cloud heights on the sequence are always Above Ground Level. Denver's lowest cloud layer is 1500′ AGL and SCaTtered. The "M"

tells you that the next layer is the "measured" ceiling, at 2500′ AGL. Here's the ceiling codes: If a minus sign is located in front of the sky coverage (ex: 30 – BKN) this indicates a "thin" layer. If the minus is with OBS, it means a partial obscuration. Remember, thin broken, thin overcast, or partially obscured sky—these do not designate a ceiling.

4. *Visibility:* In statute miles, just like your VFR minimums, and Denver reports 1 and 1/2 mile visibility. If a minus sign is associated with the obstructions to visibility, it indicates "light" and will be placed after the abbreviation for the obstruction.

5. *Millibars:* Of barometric pressure. The meteorologists work in a metric unit. Let's ignore this one an wait for inches of mercury!

6. *Temperature* Measured at the surface in degrees Fahrenheit. Always compare this with the:

7. *Dewpoint:* To see if there is a chance of visible moisture. You can easily forecast the base of the clouds by dividing the temp/dewpoint spread by 4.4, then multiplying the result by 1000. Your answer is the cloud height above the altitude where the temp/dewpoint was measured.

8. *Wind:* First two digits are the direction the wind is from rounded off to the nearest 10 degrees and stated in degrees from True North, followed by the velocity in knots. Denver reports a south wind at 07 knots. A "G" here would mean gusting.

9. *Altimeter setting:* The barometric pressure corrected to sea level, with the first digit left off. Just prefix it with a 2 or a 3 as appropriate (remember that the standard is 29.92), then add a decimal for the last two digits. Denver's got to be 29.93, as 39.93 would be much too high.

10. *Remarks:* Most anything can end the sequence—this one is "Breaks IN the OVer-Cast."

To summarize, I would read Denver's weather as: "2300Z observation, 1500 scattered, measured 2500 overcast ceiling, visibility 1 and 1/2 miles in light rain and smoke, temp 58, dewpoint 56, winds 180 at 7, altimeter 29.93."

Obviously this takes a bit of practice, so write these out and check with your ground instructor:

RAP SA 1251 250 –SCT 15 042/36/21/1909/959

BFF SA 1253 200 –SCT 25 088/27/19/0805/974

LAR SA 1252 E120 BKN 250 OVC 15 08/35/22/2313/983

SLC SA 1251 E110 OVC 30 079/53/28/1916G24/981/VIRGA ALQDS

RAP SA 1352 250 –SCT 35 028/45/20/2415/958

BFF SA 1351 80 SCT 200 –BKN 25 090/29/20/0000/975

LAR SA 1355 50 SCT E100 BKN 250 OVC 40 090/33/23/2310/983

The Surface Analysis Chart

This thing is much too detailed for a pilot to spend a great deal of time on, but take a look to get your own "big picture" of the pressure systems, fronts, and wind as of 1300Z (Figs. 4-7, 4-8).

Can you see how a pilot getting the weather for a trip along the Appalachian Mountains might want to check a little further? Which way is that front moving? Trace out the flow of wind around the low in northern Louisiana. Is it moist or dry air following the front? Is it a cold front or a warm front? Can you see how a pilot on a flight from New York to San Francisco could get a ground speed advantage by flying between that Louisiana low and the high in Minnesota?

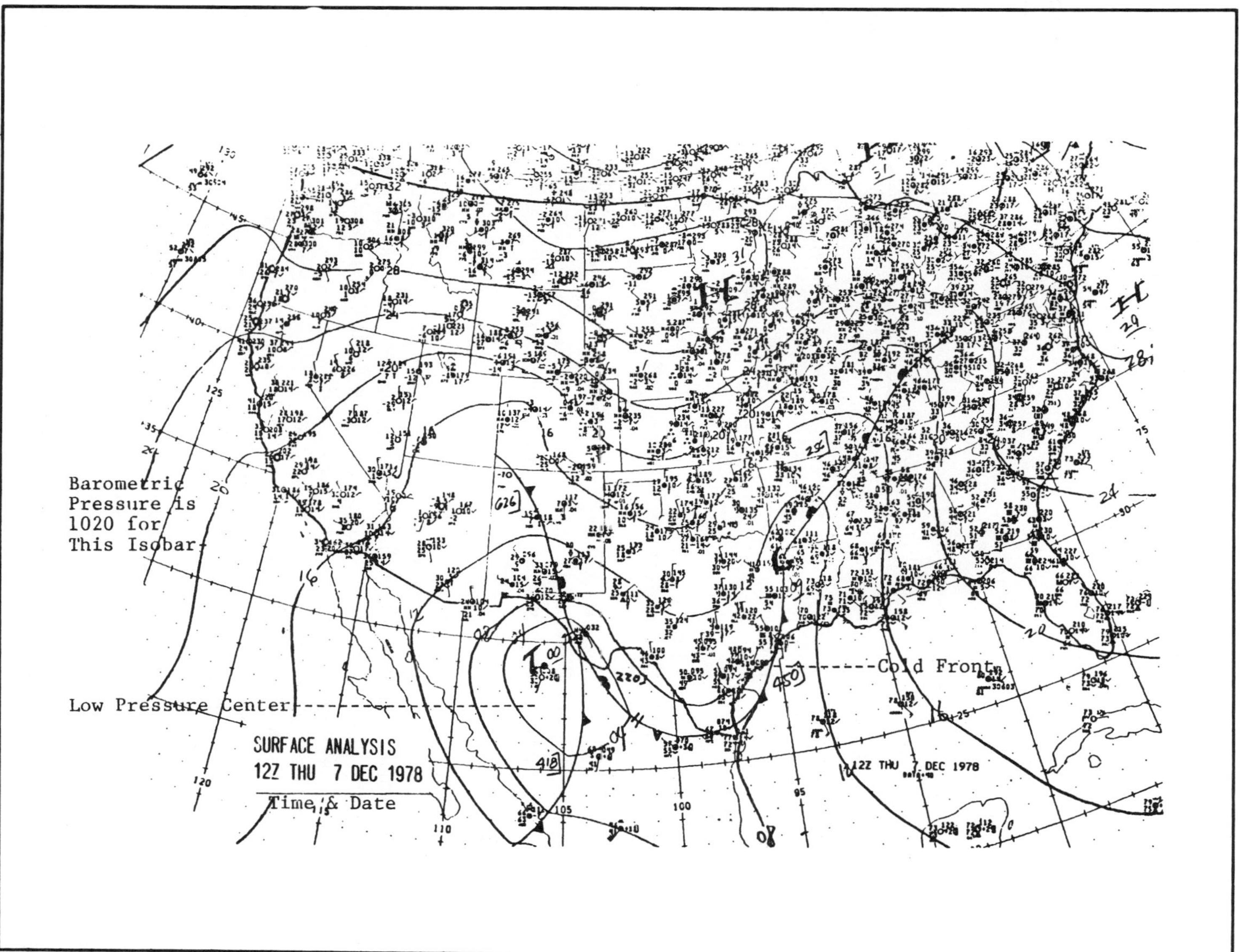

Fig. 4-7. Surface Analysis charts.

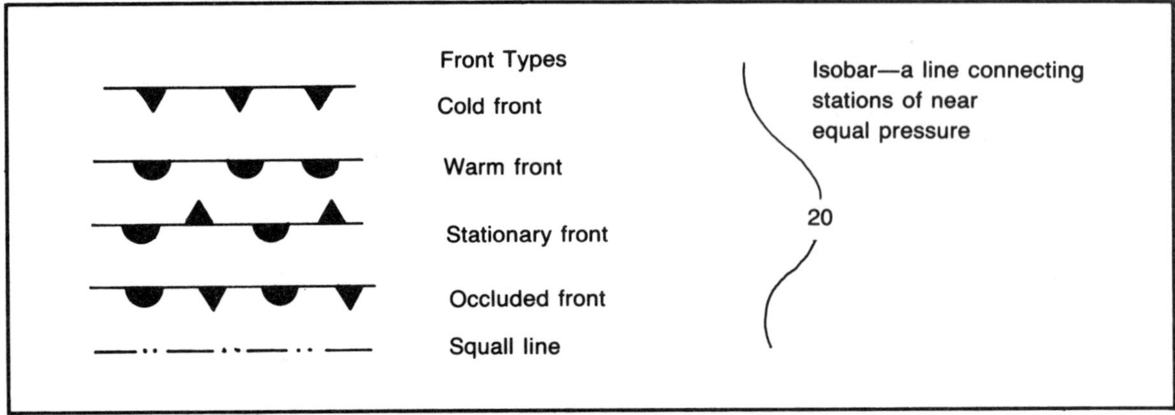

Fig. 4-8. Wind, isobars, and fronts.

The Weather Depiction Chart

Here's a chart that seems designed for pilots (Fig. 4-9). Notice it again shows the pressure systems and fronts as they were at the 1300Z observation. I'll explain the legend section in the Gulf of Mexico, as it is the best part of all. Notice the strange shape hanging over Northern Colorado. The area enclosed by the solid lines means IFR conditions, while the "scalloped" line encloses what the NWS calls "marginal VFR." You know that the weather minimums depend on altitude and controlled airspace, but the weather people use slightly different numbers, shown in the legend, to distinguish between the three classifications of pilot's weather:

- ☐ IFR means that the reported ceilings are less than 1000 feet AGL and/or visibilities are less than 3 miles.
- ☐ MVFR means that the ceilings are between 1000 and 3000 feet AGL and/or visibilities are between 3 and 5 miles.
- ☐ VFR means any area with ceilings of more than 3000 feet and visibilities of 5 or more miles.

There are other details shown on the chart near many reporting stations, but it is much easier to read the sequence report to get the exact information for any airports of particular interest.

Still want to try that flight along the Appalachians?

The Radar Summary Chart

The Radar Summary Chart is used just like the radar on television weather programs—for "seeing" heavier precipitation. In other words, radar is our best view of thunderstorms, but will not show fog or cirrus clouds. The Radar Chart is issued whenever the NWS has something interesting to show. Ask your ground instructor to show you a real one.

Perhaps our best reports of the weather are the ones the FSS collects from other pilots and broadcasts back to us that for the asking. These are called PIREPs, short for Pilot Report, and may include everything at all—no rules, symbols, or numbers to memorize! When you're cruising down an airway, the only person in the universe that knows your weather better than you do is the pilot ten minutes ahead of you at the same altitude! The FSS always encourages us to give Flight Watch a

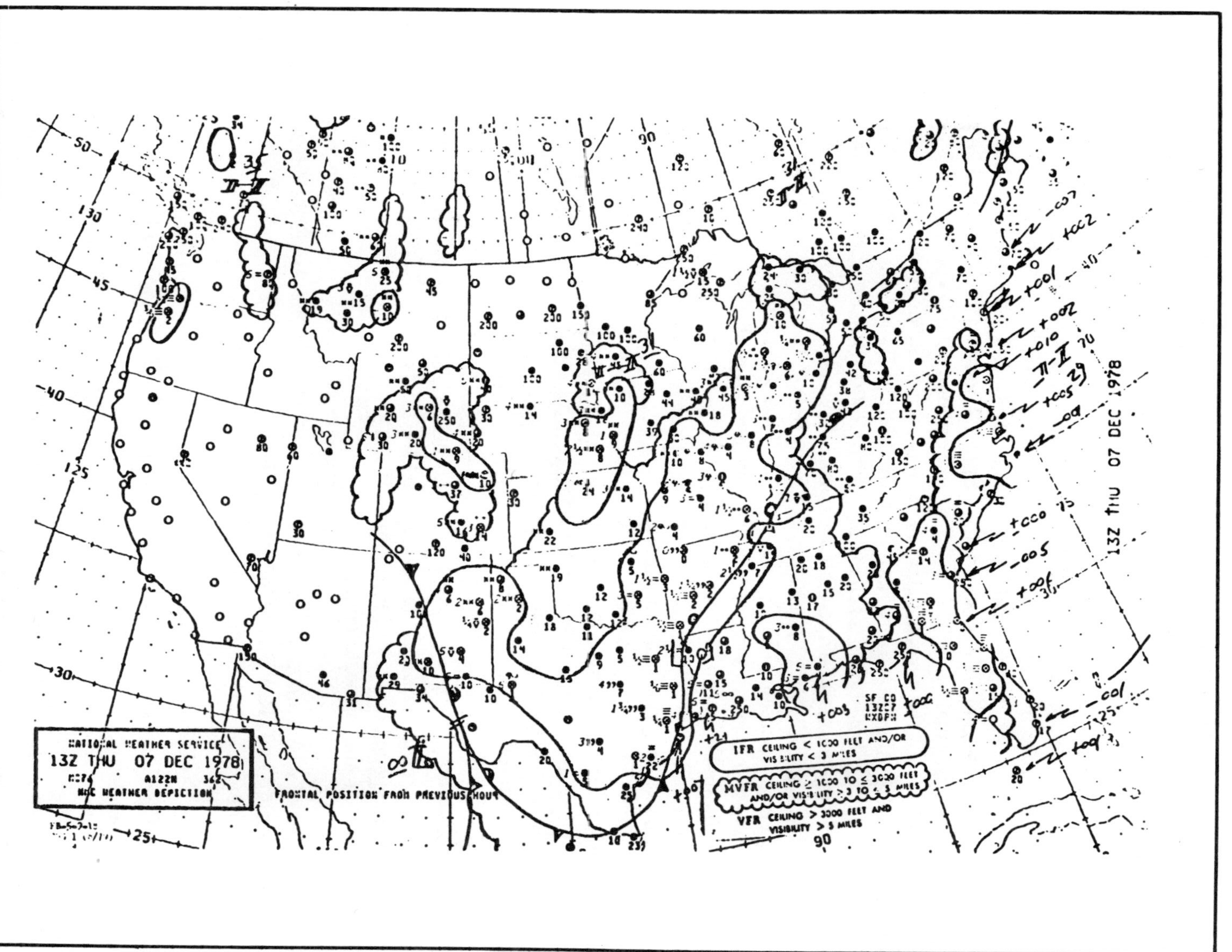

Fig. 4-9. Weather depiction chart.

PIREP, so even if you feel that there is nothing happening, give them your location, the outside air temperature, and a "smooth ride at X thousand feet." The pilot who is experiencing turbulence a few thousand feet above or below you will certainly appreciate your PIREP. Oh yes, always include your aircraft type when reporting turbulence, since moderate turbulence reported by a Boeing 727 is severe to extreme for a little Cessna!

And now we go into the forecast part of our discussion. Remember that from here on this is not *real* weather, but merely a scientific guess. The NWS only claims an 87 percent accuracy rate but you don't want to get in trouble with the other 13 percent. On the other hand, the FSS weather briefer tends to "cover himself and the federal government" by making the forecast sound scary. Of course, it's a foolish pilot who assumes the weather will be better than predicted, but a forecast of "VFR not recommended" at the destination does not restrict an efficient pilot from going out to "take a look" and perhaps settling for an alternate near his destination.

Terminal Forecasts

Selected high-traffic terminals (airports) are furnished with special forecasts for the immediate vicinity of the airport, covering a period of 18 hours with an additional six hour "categorial outlook" which amounts to the best guess of whether it'll be IFR, VFR, or MVFR. These are read almost like sequence reports, so let's try one:

SLC 251010 C5X1/2S-3325. 14Z OCNL
COX+BS3020. 19Z C150VC
3S-2915. 23Z C25BKN 10. 4Z VFR . . .

First off, the FT is for Salt Lake City and is a forecast for the 25th day of the month from 10000Z to 1000Z on the 26th. Notice how the forecast is broken up into sentences with a Zulu time for each segment. From 1000Z to 1400Z they forecast ceilings of 500 feet (AGL) and obscured (X), 1/2 mile visibility in light snow, winds 330 degrees at 25 knots. From 1400Z to 1900Z it really gets nasty: OCcasioNal Ceilings of 0 and obscured by heavy Blowing Snow. Then by 1900Z it should be getting better: ceilings 1500 OVerCast and 3 miles in light snow, wind 290 degrees at 15. From 2300Z on the 25th to 0400Z the 26th the field is VFR with 2500 Broken Ceilings and 10 miles visibility. The outlook is for VFR to continue from 0400Z to 1000Z the 26th.

That was really tough one. How about this one for Memphis?

MEM FT 132222 40SCT 300SCT 40. 02Z
CLR. 16Z VFR . . .

You see, when the weather's good, they're pretty easy. Here's a few of the FTs from the FAA test:

GAG FT 011515 100 SCT 250 SCT 2610. 16Z 60 SCT C100 BKN 3315G22 CHC C50 BKN 5TRW. 01Z 250 SCT 3515G25. 09Z VFR WIND. .

HBR FT 011515 C120 BKN 250 BKN 3010. 17Z 100 SCT C250 BNK 3215G25 CHC C30 BKN 3TRW. 00Z 250 SCT 3515G25. 09Z VFR WIND. .

MLC FT 011515 C20 BKN 1815 OCNL SCT. 20Z C30 BNK 1815G22 CHC C20 BNK 1TRW. 03Z C30 BNK 2015 CHC C7 X 1/2TRW+G40. 09Z MVFR CIG TRW..

OKC FT 011515 C12 BKN 1815G28 LWR BKN V SCT. 18Z C30 BNK 250 BNK

2315G25 LWR BKN OCNL SCT CHC C7 X 1/2TRW + G40. 21Z CFP 100 SCT C250 BKN 3315G25 CHC C30 BNK 5TRW-. 02Z 100 SCT 250 SCT 3515G25. 09Z VFR WIND..

PNC FT 011515 C100 BKN 250 BKN 250 BKN 1810. 16Z CFP 20 SCT C100 BNK 3115 SCT V BKN. 00Z 250 SCT 3515G25. 09Z VFR WIND.

TUL FT 011515 C20 BNK 1915G22. 19Z C30 BKN 1815G25 CHC 3TRW. 09Z VFR WIND..

Area Forecasts

Next on the agenda is the Area Forecast, which covers an area of several states (Fig. 4-10). It is primarily used to determine forecast enroute weather and also covers airports where a Terminal Forecast is not reported.

Need some help? The Area Forecast (FA) was issued on the 1st day of the month at 1240 Zulu, and is from the Dallas-Ft. Worth station, valid from 1300Z to 0700Z on the 2nd, with an additional outlook (guess) for the period from 0700Z to 1900Z. For New Mexico, Oklahoma, Texas and its coastal waters. Heights are MSL unless noted. Thunderstorms will imply possibly severe or greater turbulence, severe icing, and low-level wind shear.

Flight Precautions: For southwestern TX west of the Pecos River and NM for occasional moderate TURBC below 15000 with strong up-and-down drafts in the vicinity of the mountains for south central TX and southeastern TX . . . patchy ceilings at or below 1000 and visibility below 3 miles improving after 15Z. For TX and OK along and within 100 MI of the cold front . . . OCNL MDT TURBC BLO 10000 and W of front with low level wind shear.

Synopsis: CDFNT in the VCNTY of Wichita to Lubbock to Hobart line SWward will move eastward to about (FYV?)-Big Bend LN by 07Z.

Significant Clouds and Weather: NM and portions of W TX W of (INK?)-Big Bend LN: Generally 15000-20000 scattered locally broken with lower 8000-12000 SCT northeastern NM spreading over the area AFTN and clearing after dark. Patchy 4000-5000 SCT NERN NM AFDK. Outlook VFR. OK and TX W of Wichita-Childress-(INK?) LN. CIGS 10000-15000 BNK variable to SCT and 60 MI VSBY. The 1200 foot layer will be BKN V to SCT with tops in layers to 30000 with CIGS LCLY LWR OVR eastern OK, the local CIGS 1000-1400 OVC. CIG gradually lifting to 2000-3000 BKN V to SCT and 80 MI by noon with SCT TSTMS LCLY lowering conditions BLO CIG 1000 and obscured with 2 tenths coverage by TSTMS with tops to 40000. Conditions will lift ABT 50 to 100 MI behind CDFNT to CIGS 10000 BKN. The outlook is for marginal VFR because of CIGS in the thundershowers, becoming VFR 100 MI W of the CDFNT.

Cstl Wtrs: Generally 2500 SCT. Widely SCT showers developing AFT 1800Z. Outlook for VFR becoming MVFR in the TRW by mid morning.

Icing and Freezing Level: None of consequence outside showers and TSTMS. Freezing level at 9000 in northern NM sloping to 14500 in Sern TX.

Turbulence: SWRN TX W of the Pecos RVR and NM: OCNL MDT turbc BLO 12000 with STG UDDFS in the VCNTY of the MTNS. OCNL MDT TURBC within 100 MI of the FNT BLO 10000.

Winds and Temperatures Aloft Forecasts

The winds and temperatures aloft will become familiar to you with your first cross-country work, as you try to plan for the winds

```
  FA 011240
DFW FA 011240
VALID 011300Z-020700Z
OTLK 020700Z-021900Z

NM OK TX AND CSTL WTRS...

HGTS MSL UNLESS NOTED...

TSTMS IMPLY PSBL SVR OR GTR TURBC..SVR ICG..AND LOW-LVL WIND SHEAR...

FLT PRCTN...SWRN TX W PECOS RVR AND NM...OCNL MDT TURBC BLO 150 WITH
STG UDDFS VCNTY MTNS.

S CNTRL TX SERN TX...PATCHY CIGS AOB 010 AND VSBY BLO 3 MI IPVG AFT
15Z.

TX AND OK ALG AND WITHIN 100 MI OF CDFNT...OCNL MDT TURBC BLO 100 AND
W OF FNT LLWS.

SYNS... CDFNT VCNTY ICT-LBB-HOB LN SWWD WL MOV EWD TO ABT
FYV-BIG BEND LN BY 07Z.

SIGCLD AND WX...
NM AND PTN OF W TX W OF INK-BIG BEND LN...
GENLY 150-200 SCT LCLY BKN WITH LWR 80-120 SCT NERN NM SPRDG OVR AREA
AFTN AND CLRG AFDK. PATCHY 40-50 SCT NERN NM AFDK. OTLK... VFR.

OK AND TY W OF ICT-CDS-INK LN...
CIG 100-150 BKN TOPS LYRS 300. OTLK... VFR CIG ABV 100.
OK AND TX E OF ICT-CDS LN...
CIG 12-25 BKN V SCT 60 120 BKN V SCT TOPS LYRS 300 WITH CIG LCLY LWR
OVR ERN OK LCLY CIG 10-14 OVC. CIG GRDLY LFTG TO 20-30 BKN V SCT 80
BY NOON WITH SCT TSTMS LCLY LWRG CONDS BLO CIG 10 X 2TRW TOPS 400.
CONDS WL LFT ABT 50-100 MI BHD CDFNT TO CIG 100 BKN. OTLK... MVFR CIG
TRW BCMG VFR 100 MI W OF CDFNT.

CSTL WTRS...
GENLY 25 SCT. WDLY SCT SHWRS DVLPG AFT 18Z. OTLK...VFR BCMG MVFR TRW
BY MID MRNG.

ICG AND FRZLVL...NONE OF CONSEQUENCE OUTSIDE SHWRS AND TSTMS. FRZLVL
090 NRN NM SLPG TO 145 SRN TX.

TURBC...SWRN TX W PECOS RVR AND NM...OCNL MDT TURBC BLO 120 WITH STG
UDDFS VCNTY MTNS. OCNL MDT TURBC WITHIN 100 MI OF FNT BLO 100.

THIS FA ISSUANCE INCORPORATES THE FOLLOWING AIRMETS STILL IN EFFECT...
NONE.
```

Fig. 4-10. Example of an Area Forecast.

along your route of flight and find that it's usually off at least a little. Take a look at the sample in Fig. 4-11, then we'll insert that into an overall picture:

2750 + 05

The first two digits are the forecast wind direction (True) and the second two are velocity (knots), just like in the sequences (wind out of 270 degrees at 50 knots). The last group is the forecast temperature in degrees Celsius, in this case 5 degrees above freezing. A "wind/temp" group like the above is forecast for stated altitudes above selected terminals, as shown in Fig. 4. Notice the altitudes (MSL) at the top of each column, and that they don't attempt to forecast a low-level wind, such as 6,000 MSL over DEN. The winds will only be forecast for altitude levels at least 1500′ above the evaluation of the selected terminals. Temperatures will be forecast when more than 2500′AGL. A notation of 9900 means light and variable, or that the wind is less than 5 knots. If the wind is given (see STL) as 730649, we must use a different method of reading the forecast; as it is, it would read winds from 730 at 06 knots with temps at −49. Since there is no 730°, simply subtract 500 to get 230°, which is corrected, thus 230° at 06. The 06 is obviously wrong, so 100 must be added to this, giving 106 knots!

Low Level Prognostic Charts

Incidentally, the word *prognostic* means "forecast." This chart (Fig. 4-12) is the *Low Level Prognostic*, consisting of both a 12-hour (left charts) and a 24-hour (right charts) forecast that is valid for both the surface (bottom charts) and up to 24,000 feet (top charts). Study the legend of the chart for its detail.

Special Weather Advisories

These are three different unscheduled forecasts of fast-breaking weather situations intended for notification of enroute pilots of the possibility of encountering hazardous flying conditions. The three types are:

- ☐ AIRMET—advisories of potentially hazardous weather for light aircraft, such as MDT icing or TURBC and 30 knot winds.
- ☐ SIGMET—advisories of potentially hazardous weather for all aircraft such as SVR icing or TURBC, squall LNs, tornadoes, etc.
- ☐ CONVECTIVE SIGMETS—is a specialized SIGMET devoted to

FD WBC 151745
BASED ON 151200Z DATA
VALID 1600Z FOR USE 1800–0300Z. TEMPS NEG ABV 24000

FT	3000	6000	9000	12000	18000	24000	30000	34000	39000
ALS			2420	2635−08	2535−18	2444−30	245945	246755	246862
AMA		2714	2725+00	2625−04	2531−15	2542−27	265842	256352	256762
DEN			2321−04	2532−08	2434−19	2441−31	235347	236056	236262
HLC		1707−01	2113−03	2219−07	2330−17	2435−30	244145	244854	245561
MKC	0507	2006+03	2215−01	2322−06	2338−17	2348−29	236143	237252	238160
STL	2113	2325+07	2332+02	2339−04	2356−16	2373−27	239440	730649	731960

Fig. 4-11. Winds and temperatures aloft.

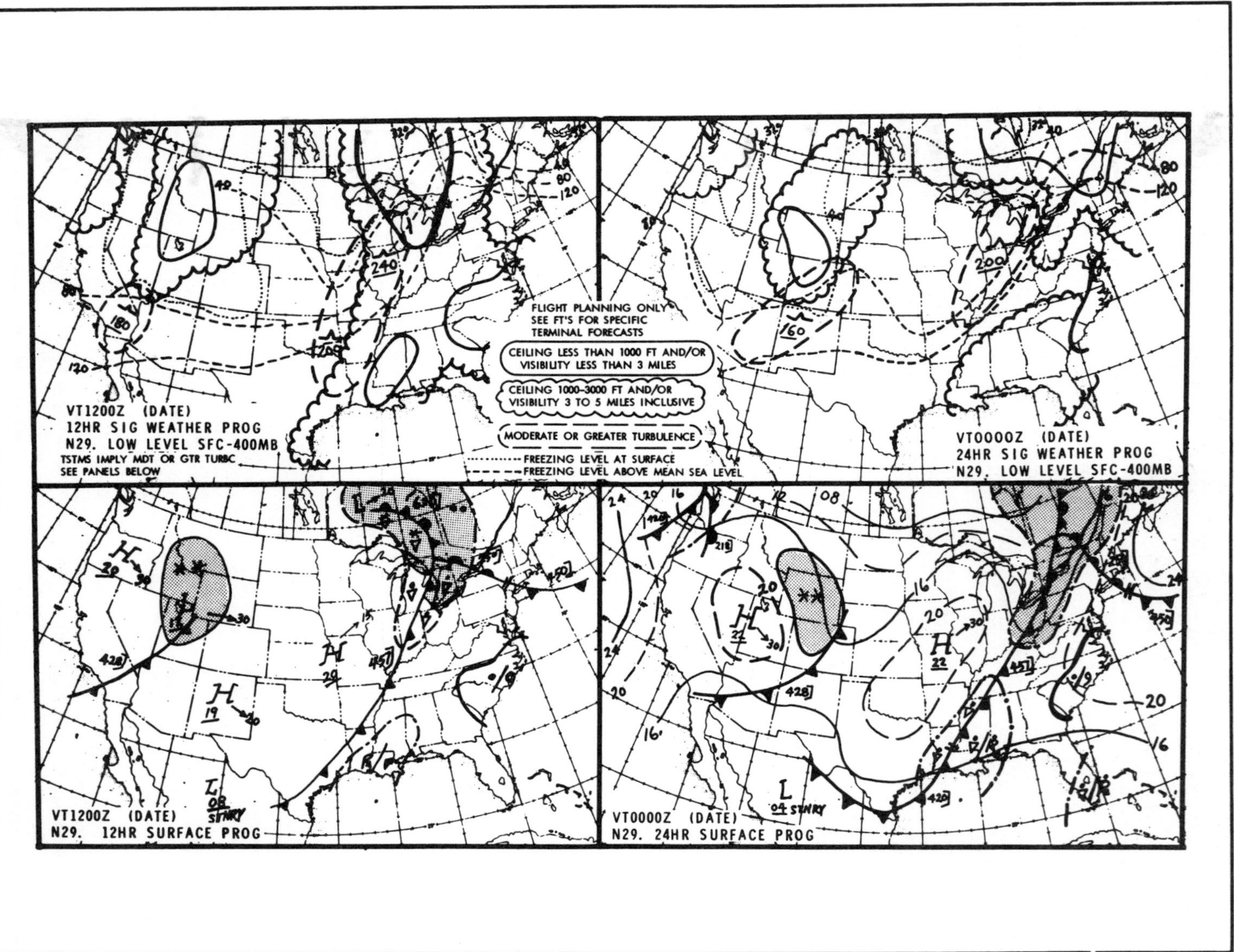

Fig. 4-12. Prognostic charts.

embedded thunderstorms, severe thunderstorms, covering more than 40 percent of the area, and hail 3/4 inch or greater in diameter.

Below are some AIRMETS samples:

AIRMET CHARLIE 1. FLT PRCTN. MTNS WRN WYO AND WRN COLO OCNL MDT TURBC BLO 180 WITH LCL STG UDDF ERN SLPGS WITH CONDS CONTG BYD 19Z.

AIRMET BRAVO 1. FLT PRCTN. MTNS NWRN WYO OBSCD IN CLDS AND SNW AOA 70 WITH CONDS SPRDG SWD AND EWD AND CONTG BYD 21Z. CONT ADVY BYD 21Z.

Chapter 5

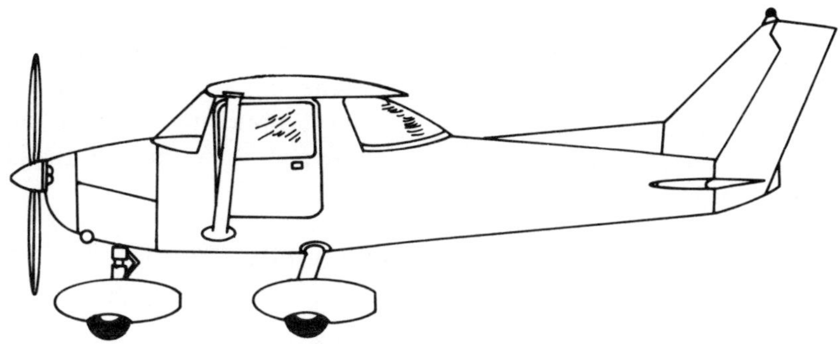

The Written Test

In the past 20 years I have seen students study, sweat, and get downright nauseous in preparation for the FAA Private Pilot Written Examination. True, the test is difficult, but thousands have taken it and thousands have passed it—therefore, it is not insurmountable!

If you look at the "philosophy" of the test, you begin to wonder why they give such an examination. It all boils down to the simple fact that, given human nature, it forces you to study. Granted, there are a few self-motivated individuals who would diligently prepare themselves for a flying career by reading anything and everything on the subject of aviation. Most of us are not that motivated; it's a lot like getting in shape. You and I both know that we need exercise and clean living, but so many other priorities take precedence over the workout sessions.

So much for the philosophy! The "bottom line" again is that you have to take the written. There are two ways you can prepare for this exam. You can run out and take a two-day "gauge" and go take the test, or you can take the subject matter in a ground school that has its curriculum spread out over several days or weeks. Which is best? That decision is up to you. Learning and comprehension vary from person to person, so if you happen to be the individual who can learn it all in two days—and remember it for later use in flight applications—then by all means enroll in the two-day course and get it out of the way. If you're not, then take it slower!

THE PRIVATE PILOT WRITTEN EXAMINATION

In this chapter we will run through all the remaining subjects to be covered on the FAA written test. You may be getting closer than you realize!

The test will be administered by an FAA-Designated Written Test Examiner. Your Private Pilot—Airplane (PA) Written Test will contain 60 questions selected for you from the nearly 1,000 questions published. The equipment needed for taking the test is a plotter and any type of navigation computer. The applicant is not permitted to use any material containing instructions related to the operation of the computer. Textbooks or notes are forbidden. The Examiner administering the test will furnish scratch paper, special pencils, and all test materials. When you show up for the "written," you must present the following:

- ☐ Proof of your eligibility to take the test (a sign-off sheet from the school).
- ☐ Proof of identity (driver's license, FAA Medical Certificate, passport, etc.).

After completing the test, your answer sheet is forwarded to Oklahoma City for scoring by electronic computers. One or two weeks thereafter, you will receive an Airman Written Test Report (Fig. 5-1) which shows your score (minimum passing is 70 percent) and lists the questions that you missed.

And if you don't pass? FAR 61.49, Retesting After Failure, states: "An applicant for a written or flight test who fails that test may not apply for retesting until after 30 days after the date he failed the test. However, in the case of his first failure he may apply for retesting before the 30 days have expired upon presenting a written statement from an authorized instructor certifying that he has given flight or ground instruction as appropriate to the applicant and finds him competent to pass the test."

BASIC AERODYNAMICS

There are four forces at work when an airplane is flying as shown in Fig. 5-2. *Thrust* is the forward force produced by the powerplant/propeller. It opposes or overcomes the force of drag. As a general rule, it is said to act parallel to the longitudinal axis. *Drag* is a retarding force caused by disruption of airflow over the wing, fuselage, and other parts. Drag opposes thrust and acts rearward parallel to the relative wind. *Weight* pulls the airplane downward because of the force of gravity, opposing lift and acting vertically downward through the airplane's center of gravity.

Bernoulli's Principle

The principle by which *lift* is produced was discovered many years ago by a scientist named Bernoulli. "The pressure of a fluid (liquid or gas) decreases at points where the speed of a fluid increases." In other words, Bernoulli found that within the same fluid (in our case, the air), a high-speed flow results in low pressure, and a low-speed flow results in high pressure. The airfoil is designed to increase the velocity of the airflow over its upper surface, thereby decreasing pressure above the airfoil (Fig. 5-3). At the same time, the impact of the air on the lower surface of the airfoil increases the pressure below. This combination of pressure decrease above the wing and the increase of pressure below the wing produces the force that we call lift.

Lift opposes the downward force of weight and acts perpendicular to the relative wind. There are four means of increasing lift:

- ☐ Increase velocity (either by increasing thrust or by diving the airplane).
- ☐ Increase the angle of attack (elevator or trim).
- ☐ Increase the surface areas of the airfoil.
- ☐ Increase the upper surface camber, or curvature.

DO NOT DESTROY THIS TEST REPORT
This Test Report must be presented for retesting or certification.

DEPARTMENT OF TRANSPORTATION FEDERAL AVIATION ADMINISTRATION

AIRMAN WRITTEN TEST REPORT (RIS: AC 8080-2)

0492 41
SSN 521-64-0486

TEST		GRADES BY SECTION							FAA OFFICE NO.	TEST DATE	EXPIRATION DATE	
TAKE NO.	TITLE *	1	2	3	4	5	6	7				
01	PA	87							RM 03	03-15-82	03-31-84	
EXPIRATION DATE (Last day of month)												

*See codes on reverse side:

MECHANICS ONLY - EXPIRATION DATE CODES
The first character designates the month; the second and third characters, the year. January through September as shown by numbers 1 through 9; October as "O"; November as "N"; December as "D"

EXAMPLES: 6 75 D 75
Month (June)
Year (1975)
Month (December)
Year (1975)

LAST NAME, FIRST MIDDLE

R

GREELEY CO 80631

NOTE: TO FIND THE SUBJECT AREA IN WHICH QUESTIONS WERE MISSED, COMPARE THE CODES SHOWN BELOW WITH THE CODED ITEMS ON THE ENCLOSED SUBJECT AREA OUTLINE.

SECTION	SUBJECT AREA CODES
1	D05 D26 K18 K21 L08 M04 P11 Q04

FRAUDULENT ALTERATION OF THIS FORM BY ANY PERSON IS A BASIS FOR SUSPENSION OR REVOCATION OF ANY CERTIFICATES OR RATINGS HELD BY THAT PERSON.

AC FORM 8080-2 (10-79)

ISSUED BY ADMINISTRATOR
FEDERAL AVIATION ADMINISTRATION

Fig. 5-1. Written test report.

That All-Important Angle of Attack

The *angle of attack* is, by definition, the acute angle between the chord of an airfoil and the direction of the relative wind. The relative wind is the airflow created by the movement of the aircraft through the air. Remember that it is *not* made up of—or affected by —the winds that are blowing in the outside world! In a sense, it's not really a wind; we're just moving the aircraft through the air. If the airplane is moving forward but settling downward (as in a glide), the relative wind is moving upward and backward!

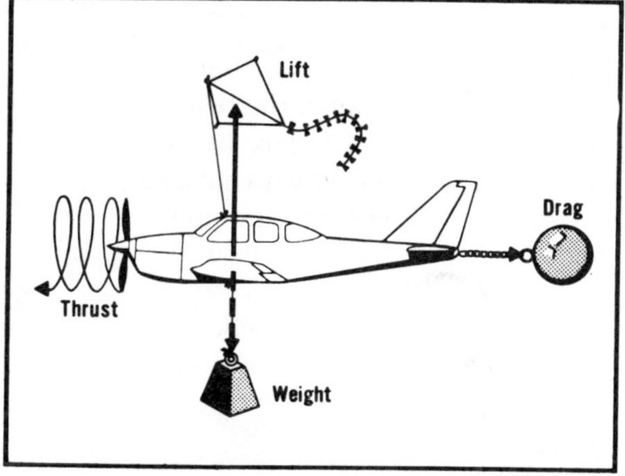

Fig. 5-2. The four forces.

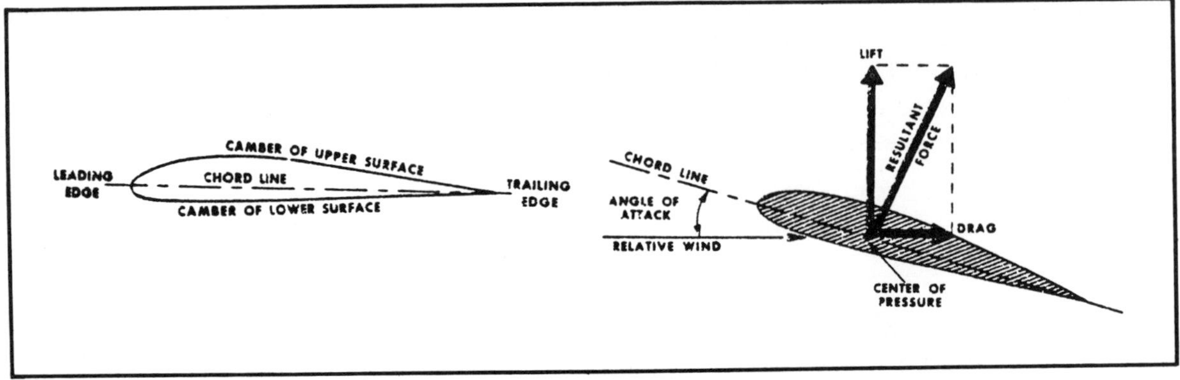

Fig. 5-3. Airfoil parts and angle of attack.

In Fig. 5-4 you can see the angle formed by the longitudinal axis of the airplane and the chord line of the wing (the angle at which the wing is attached to the fuselage), which amounts to a "built-in" angle of attack.

As the angle of attack is increased, the impact pressure, or positive pressure, on the lower surface of the wing will increase. Also, the pressure above the wing will continue to decrease because the effective camber of the airfoil is increased, requiring the air to travel a greater distance in the same period of time. According to Bernoulli's principle, it must therefore travel faster, producing a greater decrease in pressure. The resulting increase in pressure differential between the top and bottom of the wing results in a greater upward force, or greater lift, as the angle of attack is increased.

The Effects of Density

Lift and drag vary directly with the density (weight per unit volume) of the air. As air density increases (low density altitude), lift and drag increase. As air density decreases (high density altitude), lift and drag decrease. This will be developed further when we get to aircraft performance.

Ground Effect

The high-speed flow over the top of the wing forces the air just aft of the trailing edge downward; we call this *downwash*. When the airplane is operating within one-half of its wing span above the ground, the downwash reacts against the ground to change the relative wind, effectively reducing drag and increasing lift. This is the "cushion of air" that you've been

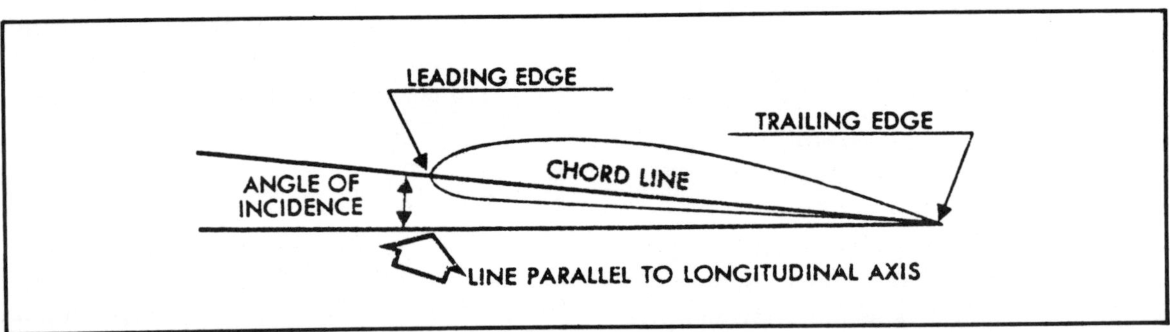

Fig. 5-4. Angle of incidence.

feeling during takeoffs and landings, and it is called *ground effect*. The unsuspecting pilot may get into problems on takeoff if he tries to climb before reaching V_y, as the loss of ground effect often causes an airplane to settle back to the surface immediately after becoming airborne.

STALLS

The angle of attack at various speeds must be such that the deflection of air is adequate for the amount of lift needed to support the aircraft's weight. If the speed is too slow, the angle of attack required will be so large that the air can no longer follow the upper camber of the wing. This results in a swirling, turbulent flow of air over the wing and "spoils" the lift. Consequently, the wing *stalls*.

On the other hand, if a pilot makes a sudden, sharp increase in the angle of attack at high speeds, the same turbulent flow along the upper surface can result. At about 15 to 20 degrees angle of attack (for light general aviation airplanes), the wing reaches its *critical angle of attack* and stalls. This can occur at *any* airspeed, in *any* attitude, with *any* power setting, and with *any* pilot, at *any* time! Remember that we stall because we exceed the critical angle of attack. This angle of attack will *always* occur at the same degree for a given airplane, even though other factors such as the airspeed may change.

SPINS

A *spin* can be described as an aggravated stall that results in what is termed *autorotation*, wherein the airplane follows a corkscrew path in a downward direction with one wing lifting and the other wing stalled. Some modern airplanes have to be forced to spin and require considerable judgment and technique to get the spin started. Then again, these same airplanes that have to be forced to spin can be accidentally put into a spin by mishandling the controls in turns, stalls, or flight at slow airspeeds.

Often a wing will drop at the beginning of a stall. When this happens, the nose will attempt to yaw in the direction of the low wing. This is where use of the rudder is important during a stall. If the nose is allowed to yaw during a stall, the airplane begins to slip in the direction of the lowered wing. This slip induces a relative wind meeting the side of the fuselage, the vertical fin, and other vertical surfaces, tending to "weathervane" the airplane into the relative wind. This accounts for the continuing yaw, or "corkscrew," which is the spin.

Spin Recovery

Any time a spin is encountered, regardless of the conditions, the normal spin recovery sequence should be used:

- ☐ Retard power.
- ☐ Apply opposite rudder to slow rotation.
- ☐ Apply positive forward elevator movement to break the stall.
- ☐ Neutralize rudder as spinning stops.
- ☐ Gently pull out of the resulting dive.

Slow and overly cautious control movements during spin recovery must be avoided. It has been found that such movements may result in the airplane continuing to spin, even with full opposite controls Brisk and positive operation, on the other hand, results in a more positive recovery.

AIRFOIL SHAPE AND FLAPS

As the upper curvature, or camber, of an airfoil is increased (up to a certain point), the lift produced by the airfoil increases. Wings that are capable of producing high lift at low

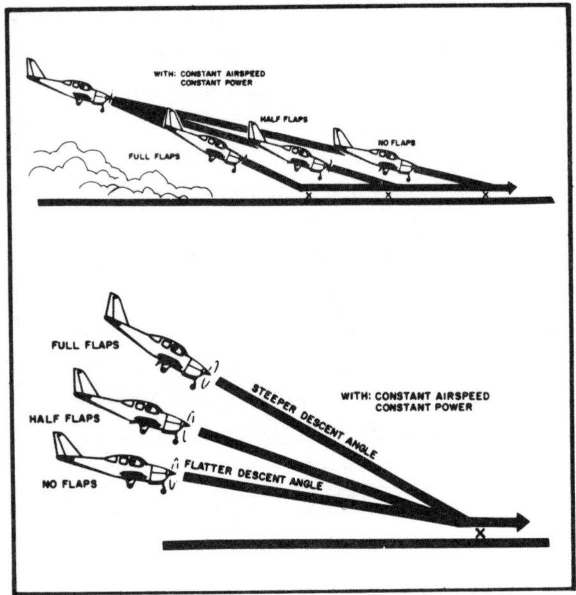

Fig. 5-5. Effect of flaps on landing point.

airspeeds have a large curvature on the upper surface and a concave lower surface. This "birdlike" shape is approximated by the use of wing flaps, thus increasing the lift (and drag) of the wing in the lower speed ranges.

The flaps are movable panels on the inboard trailing edges of the wings. They are hinged for extension into the airflow beneath the wings to increase lift and drag. Cessna trainers use a *Fowler-type* flap, which extends rearward as well as downward, thereby increasing wing area at the same time.

The purpose of flaps is to permit a slower airspeed and a steeper angle of descent during the landing approach. This will allow the pilot to land precisely on his chosen point of the runway (Fig. 5-5).

LOAD FACTORS

Load factor is the ratio of the load supported by the wings of the airplane compared to the actual weight of the airplane and its contents. This means that it is the actual load supported by the wings at any given time, divided by the weight of the airplane and its contents. At a load factor of 2, the wings support twice the weight of the airplane and its contents. In pilot talk, this is spoken of as *2Gs*.

Figure 5-6 shows very clearly that load factor is strictly a matter of bank angle. (There are possible exception, but the chart assumes coordinated flight at a constant altitude.) As the bank angle increases, the load factor also goes up. In a 60 degree bank, the load factor is 2—or 2Gs.

The Utility Category airplane (trainer) can withstand 4.4Gs without permanent structural damage. The Normal Categories are certified up to 3.8Gs.

WEIGHT AND BALANCE

All airplanes are designed for certain weight and balance conditions. Responsibility for making sure that the weight and balance limitations are met before takeoff rests with

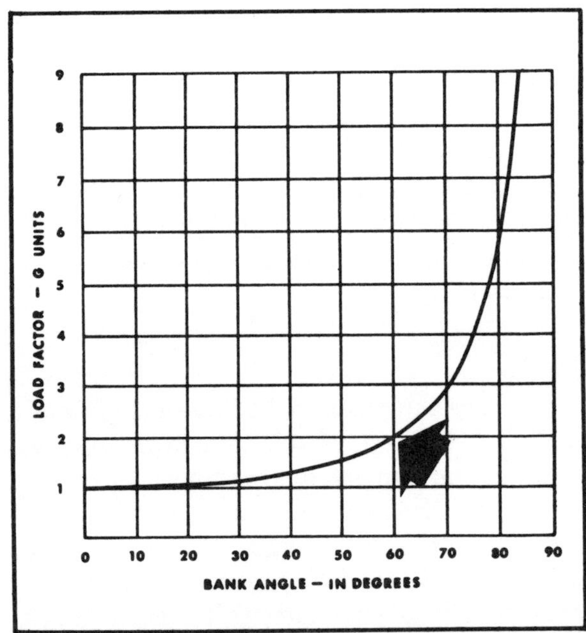

Fig. 5-6. Load factor chart.

the pilot. Three kinds of weight must be considered in the loading of every aircraft. These are empty weight, useful load, and max gross weight. *Empty weight* is the weight of the basic airplane—the structure, powerplant, fixed equipment, any fixed ballast, the unusable fuel supply, and hydraulic fluid. The modern tendency is to include the weight of the oil in what we call the "standard" empty weight. Empty weight is taken from the paperwork aboard each aircraft, and the pilot should always check to see whether the oil is included in that figure.

Useful load is the weight of pilot, passengers, baggage, and usable fuel. The empty weight plus the useful load is the gross weight of the airplane at takeoff. When an airplane is carrying the maximum load for which it was certificated by the FAA, it is at its *maximum gross weight*.

Now imagine, if you will, an airplane hung from the ceiling of a building, such as a hanger; it is hanging level (Fig. 5-7). It's not hard to see how the addition of a weight would cause the attitude to change. If you added weight, such as someone climbing into the rear of the aircraft, the plane would drop its talk. Now, to get the thing back level (with the person in

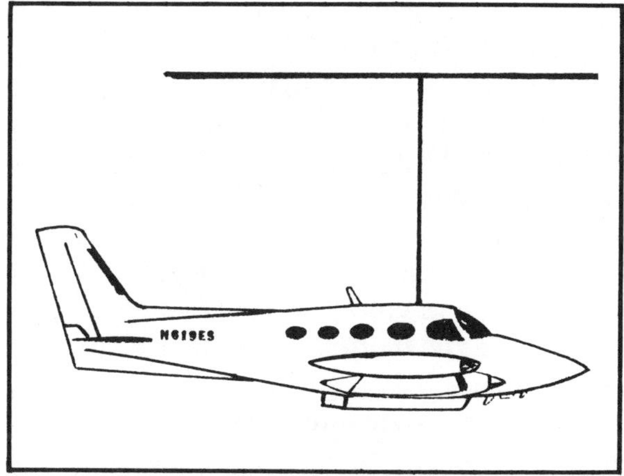

Fig. 5-7. Center of gravity by suspension.

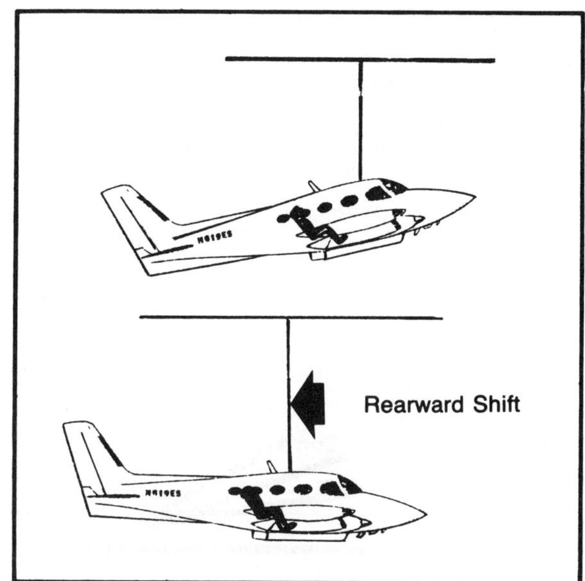

Fig. 5-8. Rearward weight shifts.

the rear seat) would require a movement of the "hook" holding the aircraft to the cable. To bring the airplane back level, the cable would have to be moved aft. Therefore, the center of gravity of the aircraft has shifted to the rear (Fig. 5-8).

The *center of gravity* (CG) is simply that point at which all the weight of the aircraft can be considered to be concentrated. It is a "balance point," and its precise location can affect the inflight performance and handling characteristics (stability). Fortunately, we do not have to "hang" the airplane to determine the center of gravity.

In Fig. 5-9 you will see the *datum line*. This is a mathematical point of reference from which calculations are made to determine the center of gravity location after the weights have been added, or to measure the *arm* of any particular point we happen to be considering, such as a baggage compartment. In other words, the arm is the distance from the datum to the weight, shown in this case as the center of gravity. The weight multiplied by the

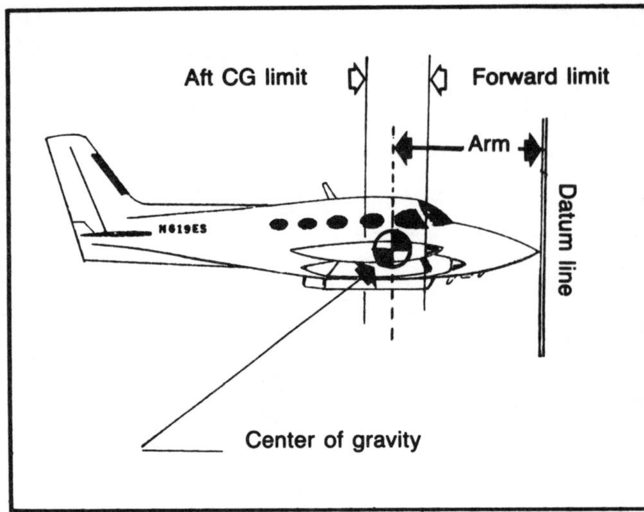

Fig. 5-9. CG location and limits.

arm gives a product called the *moment*, which is simply a mathematic expression of weight (in pounds) times distance (inches of arm) with the result expressed in *pound/inches* or *inch/pounds* of moment.

WAM!

The *forward and AFT CG limits* are established by the manufacturer and "legalized" by the FAA (FAR 91.31) to keep us from flying anything that the factory test pilot didn't have the guts to fly! For example, loading an aircraft rear of the aft CG limit is known to cause difficult stall recovery. These forward and aft CG limitations are also expressed in terms of inches aft of datum.

But remember, any weight times the distance from the datum (arm) is the moment. We call it the "WAM" formula:

$$W \times A = M$$

Now, the bottom line is to find out where the center of gravity will be when we are all through loading the airplane—and of course, the "where" implies a location that is a given distance from datum. So the problem is to find the "arm" of the loaded aircraft which is the center of gravity.

The first step is to collect the weights you plan to carry and set up a loading form like the one shown in Fig. 5-10. The aircraft empty weight comes from the paperwork that's required to be on board the aircraft (and while you're at it check whether or not the oil is included in the empty weight). Remember that fuel weighs 6 pounds per gallon, and oil weighs 7.5 pounds per gallon.

The aircraft weight and balance paperwork will also show you the arms of all tanks, seats, baggage compartments, etc., perhaps by a sketch such as Fig. 5-11.

Let's assume the empty weight is given as 2000 pounds. We want to put 320 pounds of people in the front seat, and 300 pounds in the rear seat. We have 40 gallons of fuel (how many pounds is that?) and 60 pounds of baggage. And we may as well assume the oil was *not* included in empty weight and we'll carry 8 quarts (if you figure that's 60 pounds, you're wrong!).

I might mention here that Cessna likes to use the airplane's firewall as the datum, which puts the oil ahead of the datum and gives it

	Item	Arm	Moment
1.	Aircraft Empty Weight		
2.	Pilot / Copilot Weights		
3.	Rear Seat Passengers		
4.	Fuel Weight		
5.	Baggage Weight		
6.	Oil		
	Total Weight		Total Moment

Fig. 5-10. Sample loading form.

a negative arm. Would you believe a positive weight multiplied by a negative arm results in a negative moment?

So now we plug in the weights and arms as in Fig. 5-12.

The moments are now calculated by multiplying each weight by its own arm (Fig. 5-13). To keep things in order and to get maximum accuracy, I would suggest a pocket calculator such as your Jeppesen AVSTAR.

The next step is to total up the weights and the moments. Notice that the arms column has been shaded out in Fig. 5-14 because you are through with it. Remember, those individual arms were just a means to finding the moments. The rookie sometimes has a tendency to want to add the arm column and call that the CG. *Don't do it*!

We now have the grown weight of the loaded aircraft (make sure it's not over the max

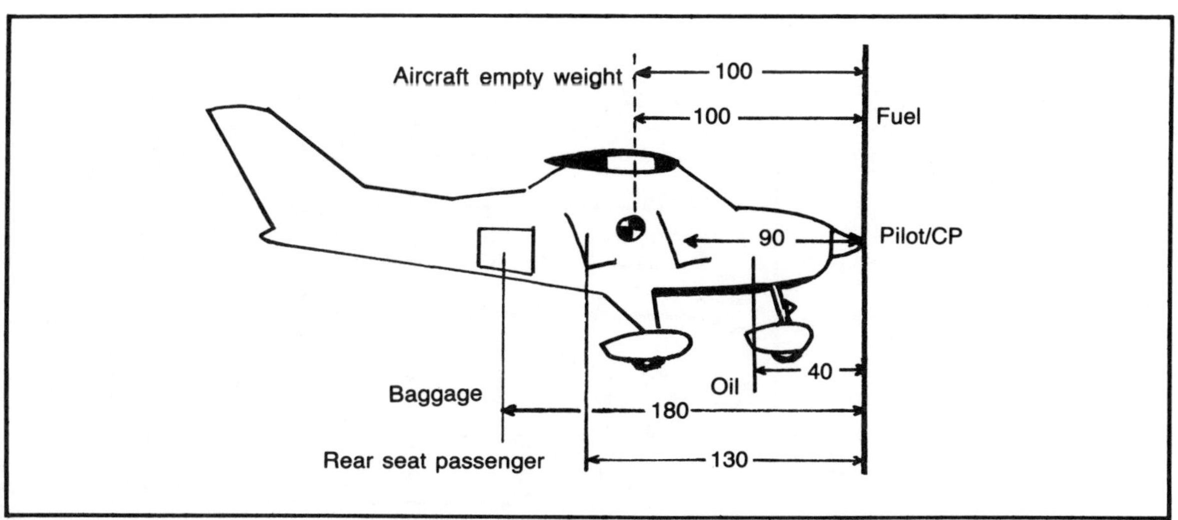

Fig. 5-11. Arms.

Weight	Arm	
2000	100	
320	90	
300	130	
240	100	
60	180	
15	40	

Fig. 5-12. Loading form with weights and arms.

W ×	A =	M
2000	100	200,000
320	90	28,800
300	130	39,000
240	100	24,000
60	180	10,800
15	40	600

Fig. 5-13. Multiply weights and arms to get moments.

ITEMS		MOMENTS
2000	100	200,000
320	90	28,800
300	130	39,000
240	100	24,000
60	180	10,800
15	40	600
2935		303,200

add add

Fig. 5-14. Add weight and moment columns.

gross), and the total moments. The last step is to divide the total moments by the weight in order to get the CG. Why? Because the "WAM" formula now becomes division instead of multiplication to solve for A instead of M. I know that 2 × 3 = 6; therefore 3 = 6/2. Or, in our problem, CG = 303,200/2935 so CG = 103.3 inches aft of the datum. And that's the loaded aircraft arm, or center of gravity, or where we can hang it from the roof of the hanger, or (if it's inside the CG limits) we can fly it!

Graphs Can Help

And now for the easy way: The graph in Fig. 5-5 solves the WAM formula for you! On the left side (vertical) you have the variable of weight, and where the weight intersects the appropriate line (arm), you can drop straight down to get the moment of that particular seat or tank along the horizontal axis of the graph.

So 200 pounds of fuel is easily found to be about 9.7 moments/ 1000 pound-inches. (That "/1000" just means we're dropping three zeros . . . the momentum is actually 9,7000 pound-inches.) Using a weight and balance form, you can just write in the weights and moments as you read them from the loading graph.

Next, to find out if your center of gravity is a go or a no-go, use the center of gravity moment envelope graph (Fig. 5-6). Plug your gross weight into the left side and plug your

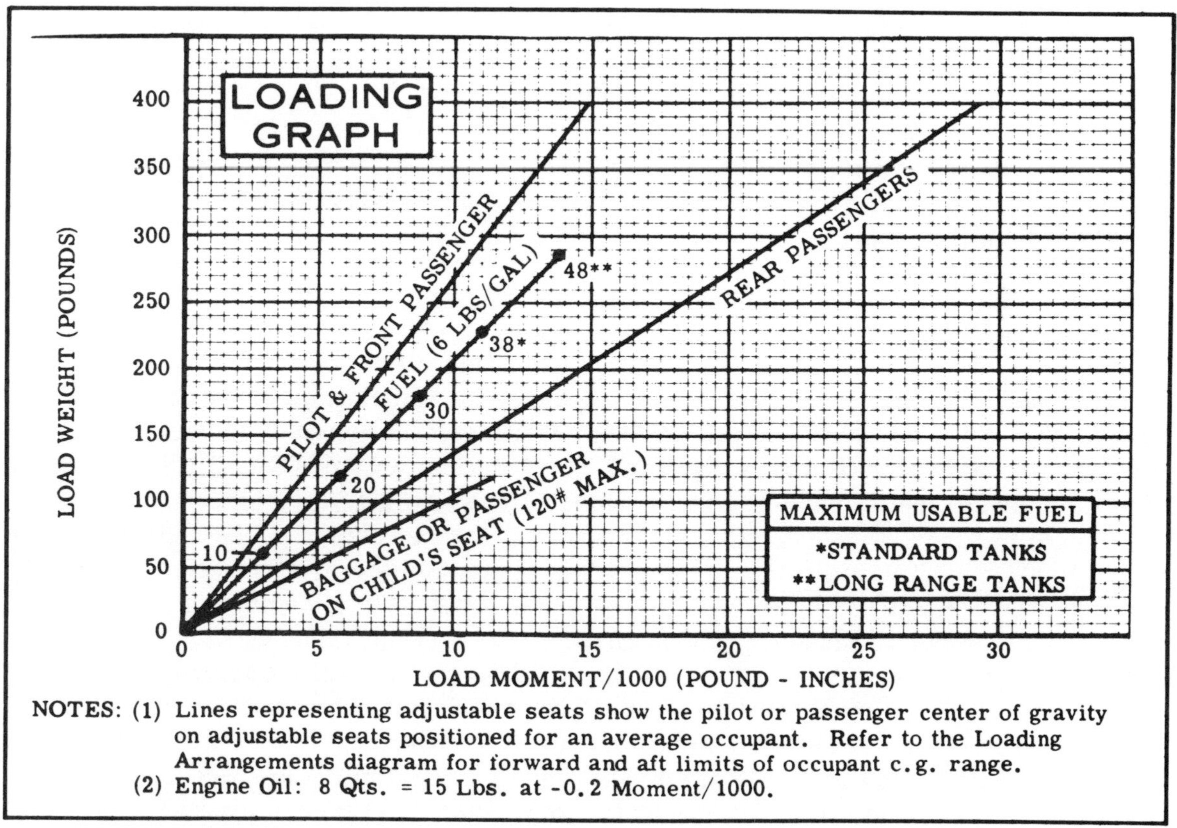

Fig. 5-15. Loading graph.

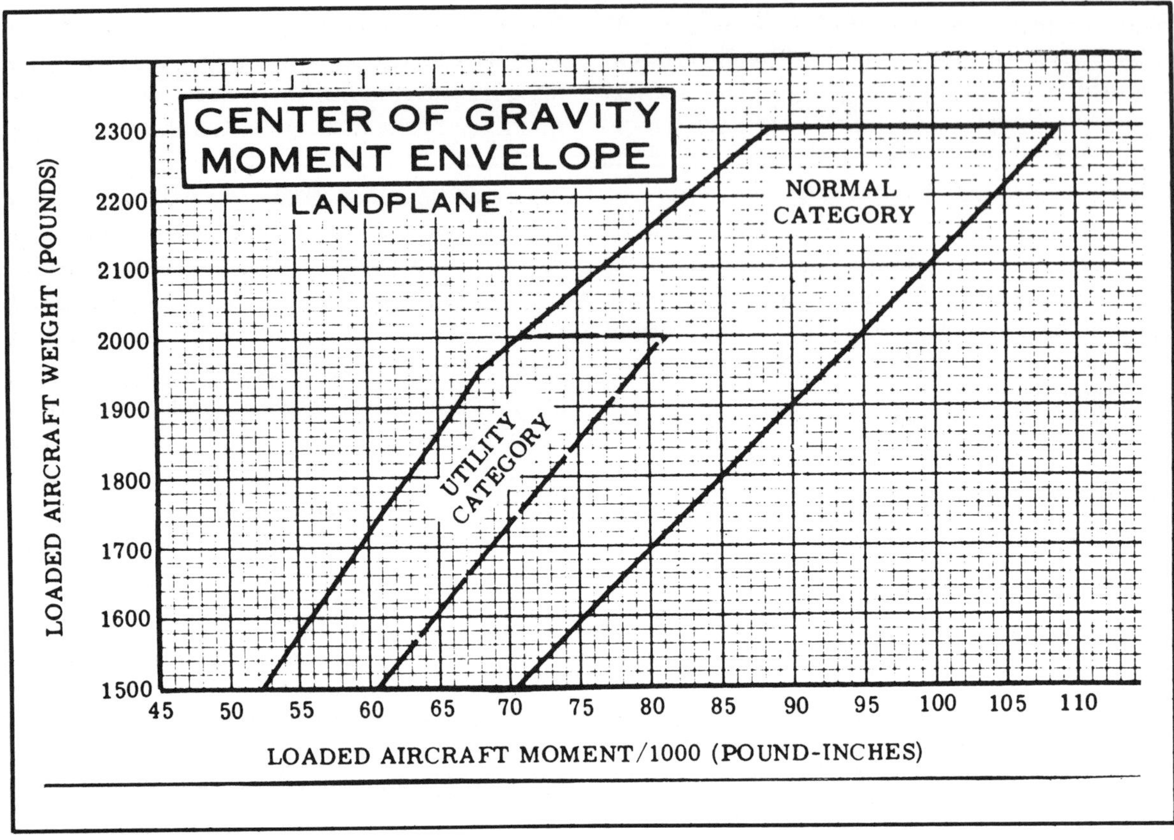

Fig. 5-16. GC/moment envelope.

total "moments/1000" into the bottom and find where the two lines intersect. If you're inside the envelope, it's legal. If you're outside the envelope, shift your loading (left is forward, right is aft). If you're above the top of the envelope, you're over max gross weight and need to remove some weight.

PERFORMANCE CHARTS

In our discussion of the aerodynamic effects of air density, it was shown that three factors came into play: pressure, temperature, and humidity. At an altitude of 18,000 feet (MSL), the density of the air is half that of sea level. Therefore, if an airplane is to maintain its lift, the velocity of the air over the wings must be increased or the angle of attack must be increased. This is why an airplane requires a longer takeoff distance at higher altitudes than under the same conditions at lower altitudes (lower pressure—higher density altitude).

Since air expands when heated, warm air is less dense than cool air (hotter temperature—higher density altitude). The pressure and temperature seem to have the most drastic effect on air density.

Because a molecule of water vapor (hydrogen and oxygen) weighs less than a molecule of dry air (nitrogen, oxygen, carbon dioxide, etc.), then it follows that moist air (high relative humidity) is less dense than dry air (high humidity—high density altitude).

All of this boils down to one major fact:

The performance of the airplane is directly related to the density altitude. The power output of the engine decreases, the propeller loses some of its effectiveness (those blades are airfoils, too), and the wing loses some of its aerodynamic efficiency anytime the density altitude is high. This is easily remembered by thinking of the pilot's worst day—low pressures (or high altitudes), high temperatures, and humid!

The Density Altitude Chart

Very important to all pilots is the density altitude chart, which cannot be overemphasized. Since you are learning to fly in the "high country," density altitude will be a deciding factor in safely getting into or out of some airports, especially in summer.

The DA chart (Fig. 5-17) allows you to input the altimeter setting and the outside air temperature at the airport to find the density altitude. This is the measurement of air density that pertains most directly to pilots. We are simply comparing the temperature and pressure to what they should be on an almost mythical Standard Day (or you could think of it as "average" or "normal"), and expressing the result in terms of feet above sea level. If that seems strange, just bear with me a minute.

First, the chart starts with the two columns of numbers on the right side. We find our altimeter setting, let's say 29.00″ Hg (low pressure day) in the left column. Just opposite, in the right column, is an 863 foot altitude correction that we must use to convert our field elevation into its pressure altitude. (Actually, if we have an altimeter handy, we can bypass this step by dialing in 29.92″ Hg and reading the result.) If we assume a 5,000 foot field elevation, it'll convert to a 5,863′ Pressure Altitude.

Now check out the slanted lines in the graph. I've "eyeballed" an extra slant-line in the chart that represents 5,863 feet, then drawn in a vertical line from an assumed outside air temperature of 70 degrees Fahrenheit.

Next a line is cast to the left (out of the graph) from the intersection point of pressure altitude and temperature. It looks to be about 7,800 feet—and that's the density altitude. Now all this means that on takeoff from *this* airport under *these* conditions, my airplane will perform as if it was taking off from an airport with a field elevation of 7,800′ MSL. that's a considerable reduction of performance. How much?

Take a look at a chart (Fig. 5-8) that visually displays the distance for the takeoff ground run and the 50′ obstacle clearance against the density altitude. Notice that the 7800′ DA from the previous problem plugged into this chart more or less goes off scale. That's what a warm, low pressure day means in colorado! By eyeball, I'd say that out ground roll will be up around 1250 feet, increased from about 1000 feet if the DA was equal to field elevation, which happens only on a "standard" day.

Takeoff Data Charts

Now, take a look at Fig. 5-19 to see how a manufacturer could present the same takeoff data in the more accurate form of a data table. First, take a look at the table and see what variables are given. You have the three gross weights, then the speed you will hold until you cross over a 50′ obstacle at the end of the runway (V_x). In each weight category line, you have the choice of three wind speeds. There are four different pressure altitudes, beginning with sea level, and with standard temperatures given. This means that by the time we correct for any non-standard temperature as directed in Note 1, we will have

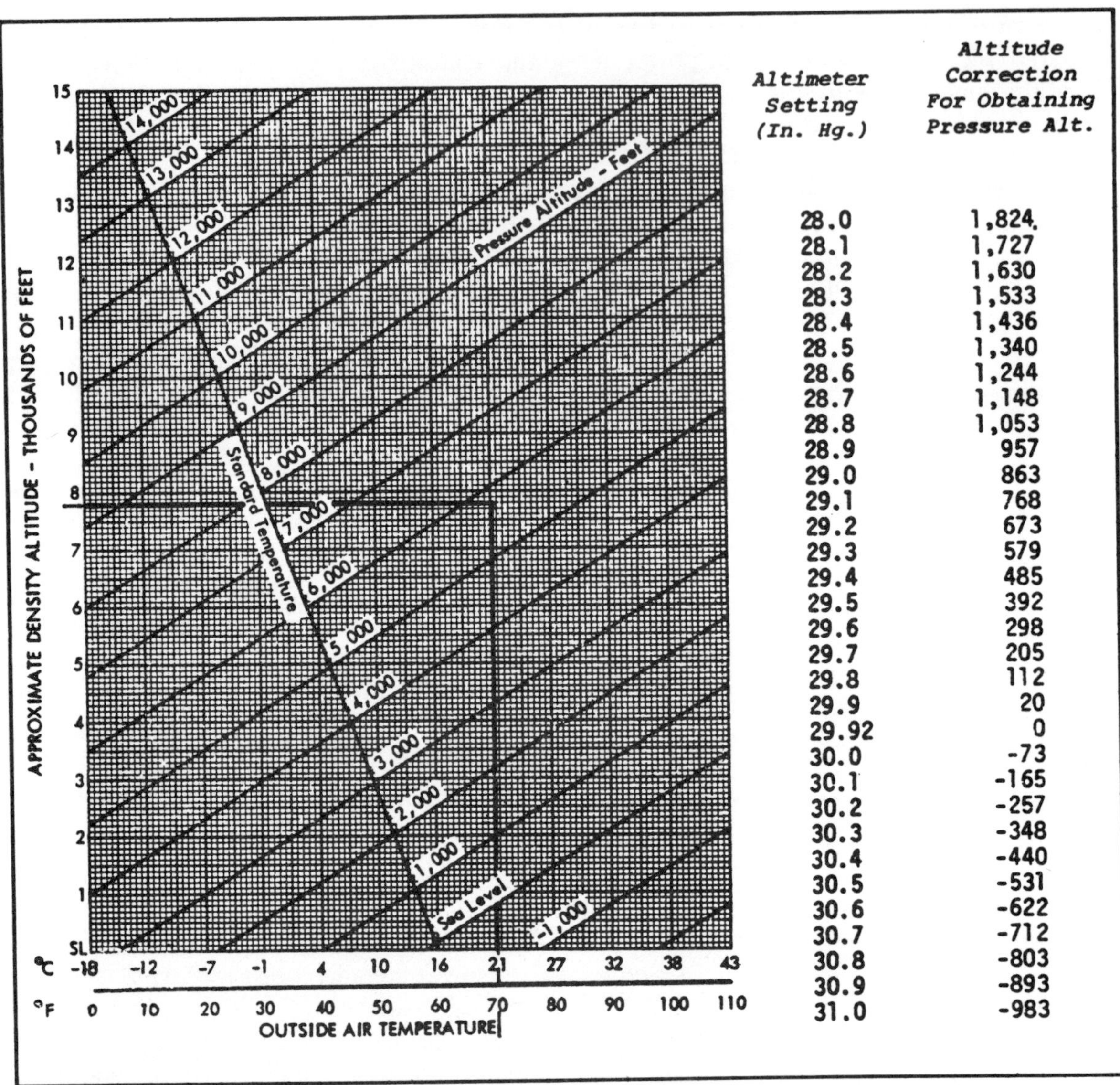

Altimeter Setting (In. Hg.)	Altitude Correction For Obtaining Pressure Alt.
28.0	1,824
28.1	1,727
28.2	1,630
28.3	1,533
28.4	1,436
28.5	1,340
28.6	1,244
28.7	1,148
28.8	1,053
28.9	957
29.0	863
29.1	768
29.2	673
29.3	579
29.4	485
29.5	392
29.6	298
29.7	205
29.8	112
29.9	20
29.92	0
30.0	-73
30.1	-165
30.2	-257
30.3	-348
30.4	-440
30.5	-531
30.6	-622
30.7	-712
30.8	-803
30.9	-893
31.0	-983

Fig. 5-17. Density altitude chart.

effectively compensated for density altitude.

For this particular aircraft, they have given the distance it takes to get off the runway, called "ground run," and the distance it takes to clear a 50′ obstacle just after takeoff. Please, don't let a chart go by you without checking it over first, carefully reading the notes usually found at the bottom. That second one is really "choice," and it means exactly what it says!

So let's say that your aircraft weighs 2000 pounds, and there's a 10 knot wind blowing right down the runway. You're at a 5,000′ airport and the temperature is a standard 41 degrees. Your takeoff run would be 645 feet and, to clear the 50′ obstacle, it would take a full 1250 feet of distance to keep the leaves

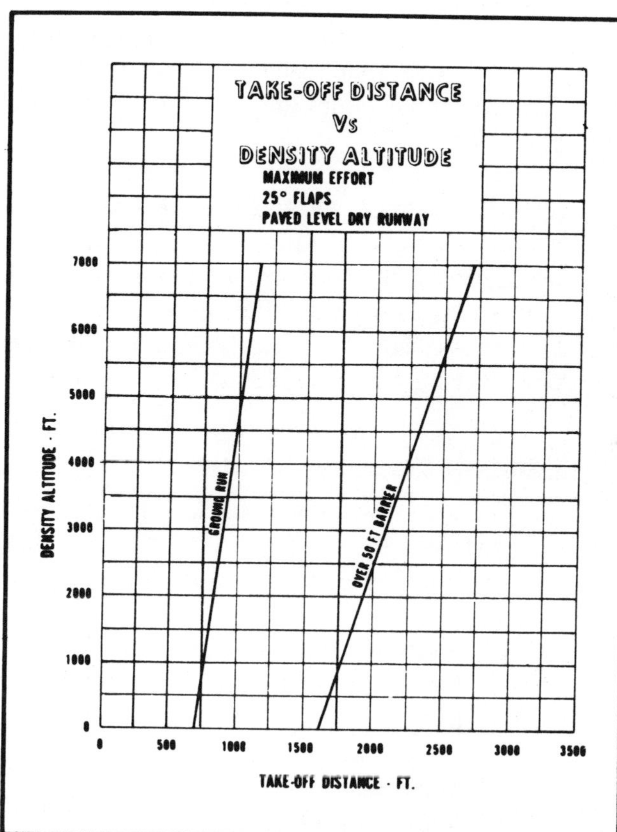

Fig. 5-18. Takeoff distance vs. density altitude chart.

TAKE-OFF DATA
TAKE-OFF DISTANCE FROM HARD SURFACE RUNWAY WITH FLAPS UP

			AT SEA LEVEL & 59°		AT 2500 FT. & 50° F		AT 5000 FT. & 41° F		AT 7500 FT. & 32° F	
GROSS WEIGHT POUNDS	IAS AT 50' MPH	HEAD WIND KNOTS	GROUND RUN	TOTAL TO CLEAR 50 FT OBS	GROUND RUN	TOTAL TO CLEAR 50 FT OBS	GROUND RUN	TOTAL TO CLEAR 50 FT OBS	GROUND RUN	TOTAL TO CLEAR 50 FT OBS
2300	68	0 10 20	865 615 405	1525 1170 850	1040 750 505	1910 1485 1100	1255 920 630	2480 1955 1480	1565 1160 810	3855 3110 2425
2000	63	0 10 20	630 435 275	1095 820 580	755 530 340	1325 1005 720	905 645 425	1625 1250 910	1120 810 595	2155 1685 1255
1700	58	0 10 20	435 290 175	780 570 385	520 355 215	920 680 470	625 430 270	1095 820 575	765 535 345	1370 1040 745

NOTES:
1. Increase distance 10% for each 25° F above standard temperature for particular altitude.
2. For operation on a dry, grass runway, increase distances (both "ground run" and "total to clear 50 ft. obstacle") by 7% of the "total to clear 50 ft. obstacle" figure.

Fig. 5-19. Takeoff data tables.

out of the landing gear.

Let's insert the "notes" and see how to mathematically do the temperature. If it was 66 degrees in the above problem, we would have to increase the distances by 10 percent. For 645 feet, we would add an additional 65 feet. For the distance to clear our 50′ obstacle, we would add 125 feet.

Rate Of Climb Charts

So now we've finally cleared the runway (and any obstacles) and we're climbing. Figure 5-20 gives us our maximum rate-of-climb under various conditions.

As I stated earlier, check the chart over first before extracting any data. We have the three sets of gross weights and four pressure altitudes. Your information obtained from the table is the rate of climb in feet per minute and the fuel used. Down at the bottom, again, please check the Notes.

Cruise Performance Charts

A cruise performance chart (Fig. 5-2) essentially tells you the power settings involved in a cruise. You can see that at 7500 feet cruising (density) altitude, using 2400 rpm, you would be using 58 percent power, truing out at 113 mph, burning 6.7 gallons per hour, and with a 38 gallon fuel capacity, have an endurance of 5.7 hours and a range (no wind) of 645 miles.

Another type of cruise performance chart (Fig. 5-22) gives you various rpm settings at chosen density altitudes, percent of brake horsepower, the true airspeed in mph, the endurance,and range based on either standard or long-range tanks. It is, essentially, like the previous performance chart.

Landing Distance Charts

We are now coming back to land and it's time to take a look at the landing distance chart (Fig. 5-23). The parameters in this chart are:

- ☐ Altitudes from sea level to 7500 feet.
- ☐ Flaps lowered to 40 degrees, power off.
- ☐ Hard surface runway.
- ☐ Zero wind.
- ☐ Notes (which see).

Don't let the single line of this small chart (Fig. 5-24) fool you . . . the conditions make all the difference. Read the notes and *all* of the parameters involved before you attempt to

MAXIMUM RATE-OF-CLIMB DATA

	AT SEA LEVEL & 59°F			AT 5000 FT. & 41°F			AT 10,000 FT. & 23°F			AT 15,000 FT. & 5°F		
GROSS WEIGHT POUNDS	IAS MPH	RATE OF CLIMB FT MIN	GAL. OF FUEL USED	IAS MPH	RATE OF CLIMB FT MIN	FROM S.L. FUEL USED	IAS MPH	RATE OF CLIMB FT MIN	FROM S.L. FUEL USED	IAS MPH	RATE OF CLIMB FT MIN	FROM S.L. FUEL USED
2300	82	645	1.0	81	435	2.6	79	230	4.8	78	22	11.5
2000	79	840	1.0	79	610	2.2	76	380	3.6	75	155	6.3
1700	77	1085	1.0	76	825	1.9	73	570	2.9	72	315	4.4

NOTES: 1. Flaps up, full throttle, mixture leaned for smooth operation above 3000 ft.
2. Fuel used includes warm up and take-off allowance.
3. For hot weather, decrease rate of climb 20 ft. min. for each 10°F above standard day temperature for particular altitude.

Fig. 5-20. Max rate of climb.

CRUISE & RANGE PERFORMANCE

Gross Weight- 2300 Lbs.
Standard Conditions
Zero Wind Lean Mixture

NOTE: Maximum cruise is normally limited to 75% power.

ALT.	RPM	% BHP	TAS MPH	GAL / HOUR	38 GAL (NO RESERVE) ENDR. HOURS	38 GAL (NO RESERVE) RANGE MILES	48 GAL (NO RESERVE) ENDR. HOURS	48 GAL (NO RESERVE) RANGE MILES
2500	2700	86	134	9.7	3.9	525	4.9	660
	2600	79	129	8.6	4.4	570	5.6	720
	2500	72	123	7.8	4.9	600	6.2	760
	2400	65	117	7.2	5.3	620	6.7	780
	2300	58	111	6.7	5.7	630	7.2	795
	2200	52	103	6.3	6.1	625	7.7	790
5000	2700	82	134	9.0	4.2	565	5.3	710
	2600	75	128	8.1	4.7	600	5.9	760
	2500	68	122	7.4	5.1	625	6.4	790
	2400	61	116	6.9	5.5	635	6.9	805
	2300	55	108	6.5	5.9	635	7.4	805
	2200	49	100	6.0	6.3	630	7.9	795
7500	2700	78	133	8.4	4.5	600	5.7	755
	2600	71	127	7.7	4.9	625	6.2	790
	2500	64	121	7.1	5.3	645	6.7	810
	2400	58	113	6.7	5.7	645	7.2	820
	2300	52	105	6.2	6.1	640	7.7	810
10,000	2650	70	129	7.6	5.0	640	6.3	810
	2600	67	125	7.3	5.2	650	6.5	820
	2500	61	118	6.9	5.5	655	7.0	830
	2400	55	110	6.4	5.9	650	7.5	825
	2300	49	100	6.0	6.3	635	8.0	800

Fig. 5-21. Cruise and range performance.

answer the FAA's questions. Try working out today's conditions on all the charts—it's worth some practice.

And, finally, one more chart for Normal Landing Distance; notice the Notes included in Fig. 5-24.

Notice that this chart presents the wind as a "down the runway" component, but you know by now that a quartering headwind is the usual landing condition. So Fig. 5-25 was developed for use with any aircraft as a means of determining what any given wind is doing to you in terms of a "down runway" headwind component, along with the crosswind component.

The Crosswind Component Chart

First, study the arcs or curved lines that represent the wind velocities. Note they start from 0 in the lower left corner and increase in all directions outwards. Now take a look at Fig. 5-26 to get your airplane oriented with the chart in Fig. 5-25.

Essentially, the headwind component (HWC) is trying to slow you down, so the value

ALT.	RPM	%BHP	TAS MPH	Endurance on 38.8 Gal (HR)	Range on 38.8 Gal (MI)	Endurance on 58.8 Gal (HR)	Range on 58.8 Gal (MI)
2500	2550 2425 2325	75 65 55	136 126 118	2.9 3.5 4.2	400 445 495	4.8 5.8 7.1	650 740 835
3500	2575 2450 2350	75 65 55	137 127 119	2.9 3.5 4.2	405 450 495	4.8 5.8 7.1	655 745 840
4500	2600 2475 2375	75 65 55	139 128 119	2.9 3.5 4.2	410 450 495	4.8 5.8 7.1	665 750 845
5500	2625 2500 2400	75 65 55	140 129 120	2.9 3.5 4.2	410 455 495	4.8 5.8 7.1	670 755 845
6500	2650 2550 2450	75 65 55	142 130 120	2.9 3.5 4.2	415 455 500	4.8 5.8 7.1	675 760 850
7500	2675 2575 2450	75 65 55	143 131 121	2.9 3.5 4.1	420 460 500	4.8 5.8 7.1	685 765 855
8500	2650 2600 2475	70 65 55	138 132 122	3.2 3.5 4.1	440 460 500	5.3 5.8 7.0	730 770 855
9500	2625 2550 2500	65 60 55	134 128 122	3.5 3.8 4.1	465 480 500	5.8 6.3 7.0	775 815 860
10500	2650 2575 2525	65 60 55	135 129 123	3.5 3.8 4.1	465 485 505	5.8 6.3 7.0	780 820 860

Fig. 5-22. Cruise performance.

derived from the vertical line should be subtracted from your airspeed, thus giving you ground speed. The CWC, or crosswind component, is trying to blow you off the runway centerline. You will have to make a correction for this. To find the value of each component requires the input of the wind direction and velocity at the airport of intended landing.

—LANDING DISTANCE— FLAPS LOWERED TO 40° - POWER OFF HARD SURFACE RUNWAY - ZERO WIND

GROSS WEIGHT LBS.	APPROACH SPEED, IAS, MPH	AT SEA LEVEL & 59° F.		AT 2500 FT. & 50° F.		AT 5000 FT. & 41° F.		AT 7500 FT. & 32° F.	
		GROUND ROLL	TOTAL TO CLEAR 50 FT. OBS	GROUND ROLL	TOTAL TO CLEAR 50 FT. OBS	GROUND ROLL	TOTAL TO CLEAR 50 FT. OBS	GROUND ROLL	TOTAL TO CLEAR 50 FT. OBS
1600	60	445	1075	470	1135	495	1195	520	1255

NOTES: 1. Decrease the distances shown by 10% for each 4 knots of headwind.
2. Increase the distance by 10% for each 60°F. temperature increase above standard.
3. For operation on a dry, grass runway, increase distances (both "ground roll" and "total to clear 50 ft. obstacle") by 20% of the "total to clear 50 ft. obstacle" figure.

Fig. 5-23. Landing distance chart.

ASSOCIATED CONDITIONS

POWER	OFF
FLAPS	35°
GEAR	DOWN
RUNWAY	PAVED, LEVEL, DRY SURFACE
WEIGHT	2750 POUNDS
APPROACH SPEED	85 MPH/74 KTS IAS

NOTES:

1. GROUND ROLL IS APPROXIMATELY 45% OF TOTAL DISTANCE OVER 50 FT. OBSTACLE
2. FOR EACH 100 LBS. BELOW 2750 LBS. REDUCE TABULATED DISTANCE BY 3% AND APPROACH SPEED BY 1 MPH.

WIND COMPONENT DOWN RUNWAY	SEA LEVEL		2000 FT		4000 FT		6000 FT		8000 FT	
KNOTS	OAT °F	TOTAL OVER 50 FT OBSTACLE FEET	OAT °F	TOTAL OVER 50 FT OBSTACLE FEET	OAT °F	TOTAL OVER 50 FT OBSTACLE FEET	OAT °F	TOTAL OVER 50 FT OBSTACLE FEET	OAT °F	TOTAL OVER 50 FT OBSTACLE FEET
0	23	1578	16	1651	9	1732	2	1820	-6	1916
	41	1624	34	1701	27	1787	20	1880	13	1983
	59	1670	52	1752	45	1842	38	1942	31	2050
	77	1717	70	1804	63	1899	56	2004	49	2118
	95	1764	88	1856	81	1956	74	2066	66	2187
15	23	1329	16	1397	9	1472	2	1555	-6	1644
	41	1372	34	1444	27	1524	20	1611	13	1707
	59	1414	52	1491	45	1575	38	1668	31	1770
	77	1458	70	1540	63	1626	56	1727	49	1833
	95	1502	88	1588	81	1682	74	1784	66	1898
30	23	1079	16	1142	9	1212	2	1289	6	1372
	41	1119	34	1186	27	1260	20	1341	13	1430
	59	1158	52	1230	45	1308	38	1395	31	1489
	77	1199	70	1275	63	1357	56	1449	49	1548
	95	1240	88	1320	81	1407	74	1502	66	1608

Fig. 5-24. Normal landing distances.

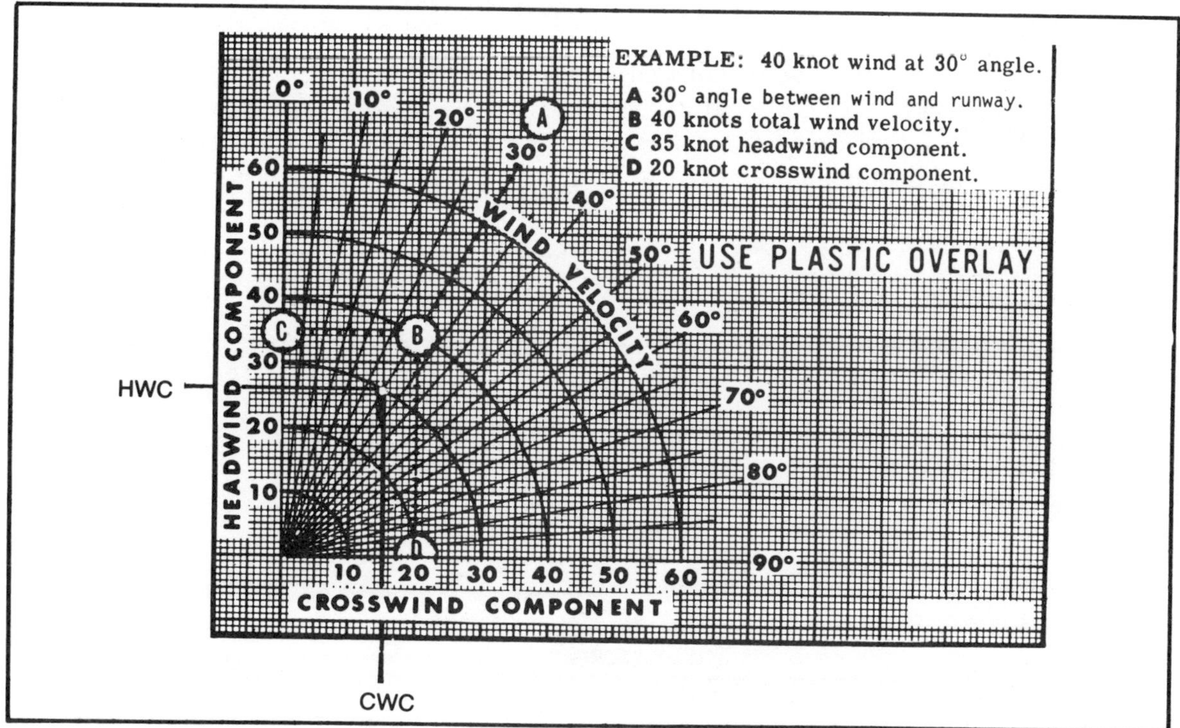

Fig. 5-25. Wind component chart.

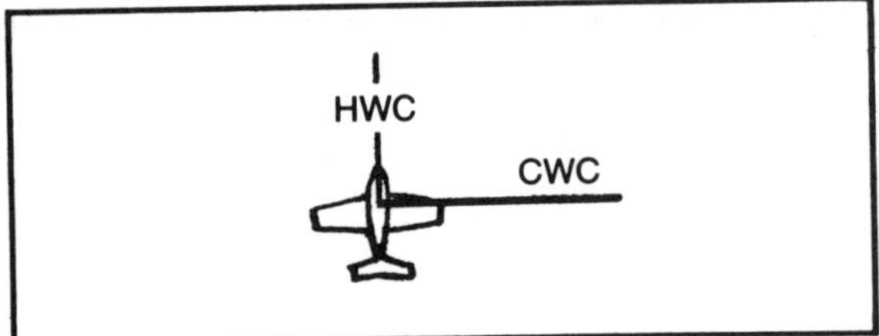

Fig. 5-26. Crosswind/headwind components.

See Fig. 5-27: Let's say you were landing on runway 09, and the wind was off your nose from 120 degrees at 30 knots. You can see that 120 – 90 = 30 degrees. So we have a 30 knot wind, 30 degrees off our nose. Using the chart, find 30 on either axis, and follow the curved line until you intersect the 30 degree slant line. Dropping straight down to the horizontal axis, you would see the CWC is 15 knots and the HWC (go left of the intersection to the vertical axis) is 26 knots.

THE INSTRUMENTS AND DEFINITIONS

In Chapter 1 we covered the basic instrument panel. On your Private Pilot Written Examination, there are several points I did not cover, so here's the details:

Kinds of Airspeeds

Indicated airspeed (IAS): The direct instrument reading the pilot obtains from the airspeed indicator, uncorrected for any of the errors.

Calibrated airspeed (CAS): This is the indicated airspeed corrected for the installation error and instrument error, the mechanical inaccuracies of the system itself. Corrected according to the chart furnished in the aircraft manual.

True airspeed (TAS): The actual speed of the aircraft through the air. CAS would be equal to TAS if we were flying in standard sea-level conditions. Since air density decreases with an increase in altitude, the airplane has to fly faster at higher altitudes to cause the same pressure difference between pilot impact pressure and static pressure. Therefore, CAS (or at least IAS) must be adjusted for pressure and temperature differences in order to obtain TAS. This correction is easily made with your flight computer.

Kinds of Altitudes

Absolute altitude: The altitude of an aircraft above the surface of the terrain over which it is flying, or AGL (Fig. 5-28).

Pressure altitude: The altitude read from the altimeter when the altimeter setting window (Kollsman window) is adjusted to 29.92″.

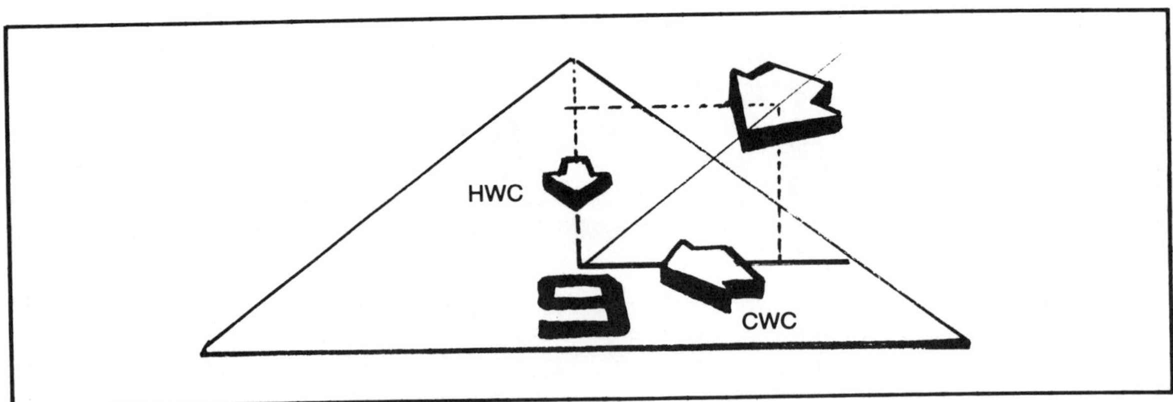

Fig. 5-27. Wind component, runway graphic display.

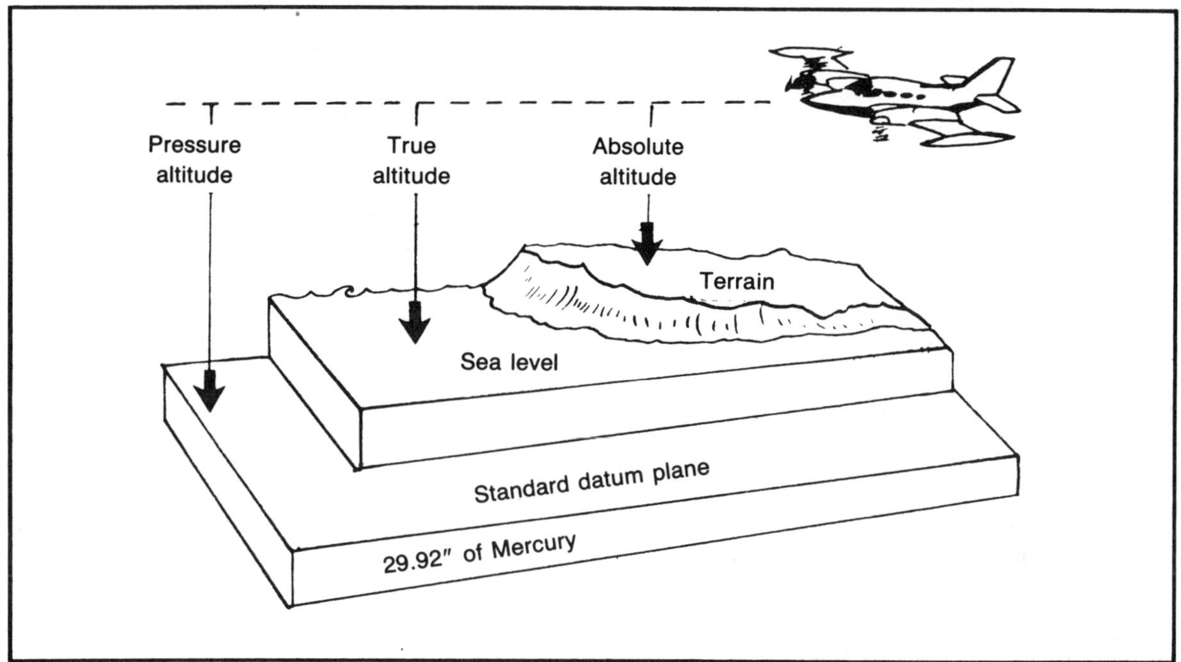

Fig. 5-28. Altitude types.

(Use for computer solutions for density altitude,the true altitude, true airspeeds, etc.)

True altitude: The actual height of the aircraft above Mean Sea Level, or MSL.

Density altitude: This altitude is pressure altitude corrected for nonstandard temperature variations. Not a "flight" altitude, but an important "performance" altitude.

Standard Conditions

Normally, as we go higher in altitude, the pressure and temperature drop. These normal drops are called the *lapse rates* and are stated in terms of 1,000-foot increments. Of course, a decrease in altitude will give corresponding increases in temperature and pressure.

The temperature lapse amounts to 3.5 degrees F. (or 2 degrees Celsius) for every 1000′. Aside form the all-important measurement of aircraft performance, the temperature lapse rate is used for such problems as measuring the height of the freezing level. (Try it, it's easy!)

Pressure, on the other hand, drops 1 inch (on a mercury barometer) for every 1000′ gain in altitude. Thus the standard pressure for Denver is 24.92 inches as opposed to 29.92 for sea level. Denver is 5000 feet and that would be a 5-inch lapse. Of course, we keep our altimeters set to sea level settings so that the outside pressure will reflect an MSL altitude.

The Compass and Its Errors

Acceleration/deceleration error can occur during airspeed changes. It is most apparent on headings of east or west. An acceleration will cause the compass card to swing towards a "north" indication while a deceleration causes a "south" indication. We remember this as the "ANDS" rule: Accelerate—north; decelerate—south.

Northerly turn error occurs when the aircraft on a northerly heading rolls into a bank. The compass card actually reverses and turns in the opposite direction at the start of the turn. Thus an aircraft heading north and rolling into a right bank (east) will cause the compass to swing first to the west. The compass card will lag behind the heading until the aircraft reaches an east or west heading.

Southerly turn error is opposite the above. An aircraft headed south and rolling into a turn will cause the compass to read in the proper direction, but more turn will be shown than is actually being accomplished. We say that the compass (like General Robert E. Lee) leads in the south, and lags in the north.

THE AIRCRAFT ENGINE

The typical lightplane engine is an internal combustion, fossil-fuel-burning, four-stroke machine capable of turning chemical energy into mechanical energy. An aircraft engine is more appropriately called a *powerplant* because it may provide power to several systems: electrical, hydraulic, pneumatic, air conditioning, etc. In the reciprocating engine (unlike the turbine), pressures from burning and expanding gases of the fuel-air mixture cause a piston to move down in an enclosed cylinder. This straight-line motion of the piston is transformed, through a connecting rod, into the rotary motion of the crankshaft, which can be splined or geared to a propeller (Fig. 5-29).

The Intake Stroke

During the *intake stroke*, Fig. 5-30, of the piston, a new fuel/air charge is drawn into the cylinder through the open intake valve. When the piston approaches the lower limit of its downward stroke, the intake valve closes and traps the gaseous mixture of air and fuel within the combustion chamber.

The Compression Stroke

Next, because both valves are closed as the

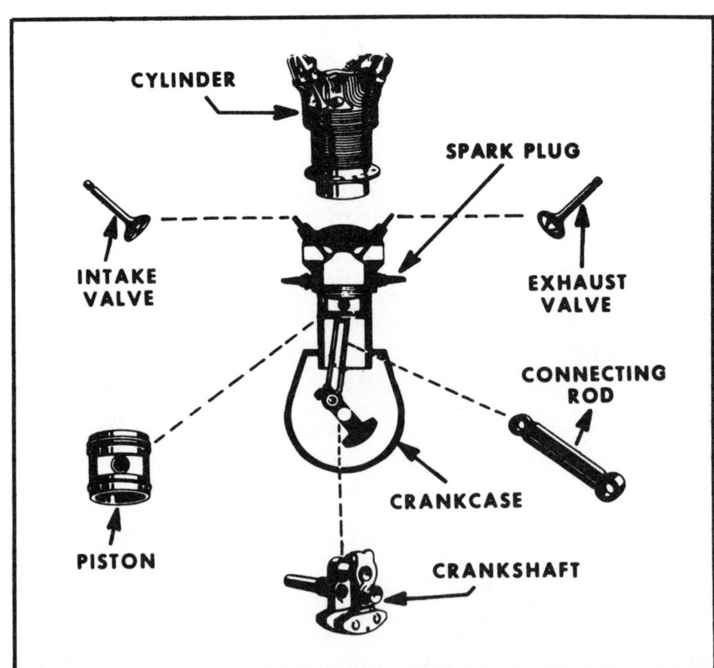

Fig. 5-29. Aircraft engine components.

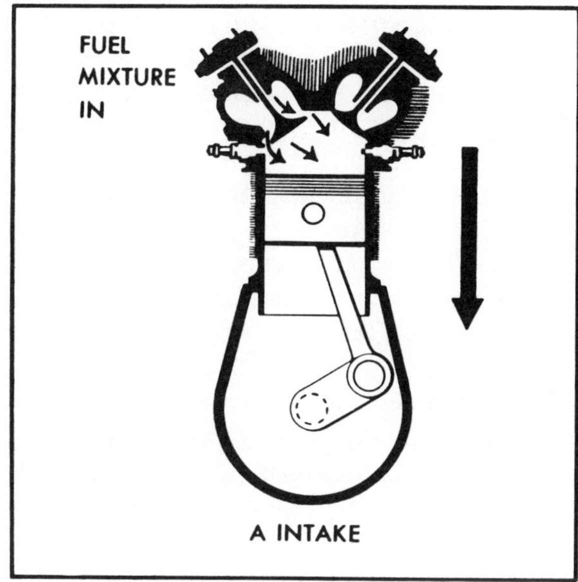

Fig. 5-30. Intake stroke.

piston moves upward, the fuel/air mixture is highly compressed between the piston and cylinder head when the upmost position is reached. This is the *compression stroke*, and the four-stroke cycle begins again as the spark plugs fire (Fig. 5-31).

The Power Stroke

At the appropriate instant (Fig. 5-32), the *power stroke* an electric spark passes across the electrodes of each spark plug in the cylinder and ignites the mixture. As the fuel/air mixture burns, temperature and pressure within the cylinder rise rapidly. The gaseous mixture, expanding as it burns, forces the piston downward and causes it to deliver mechanical energy to the crankshaft. Both valves are closed during the power stroke.

The Exhaust Stroke

The energy delivered to the crank during the power stroke causes the crankshaft to rotate on its bearings. Continued rotation causes the piston to move upward again on the *exhaust stroke* (Fig. 5-33). The exhaust valve that opened during the latter part of the

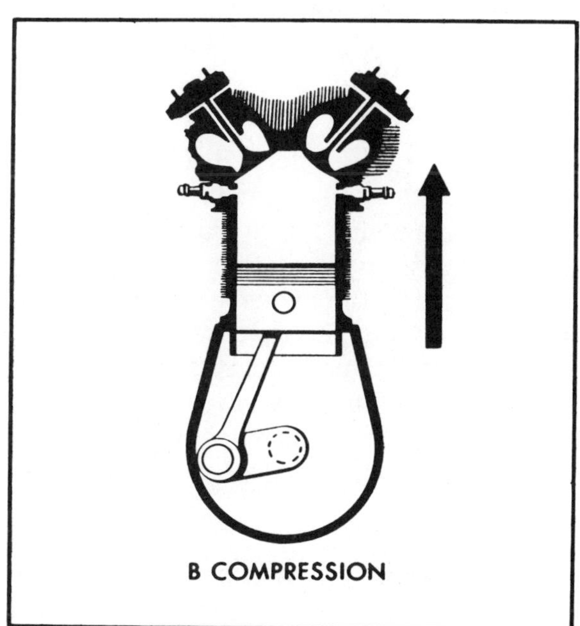

Fig. 5-31. Compression stroke.

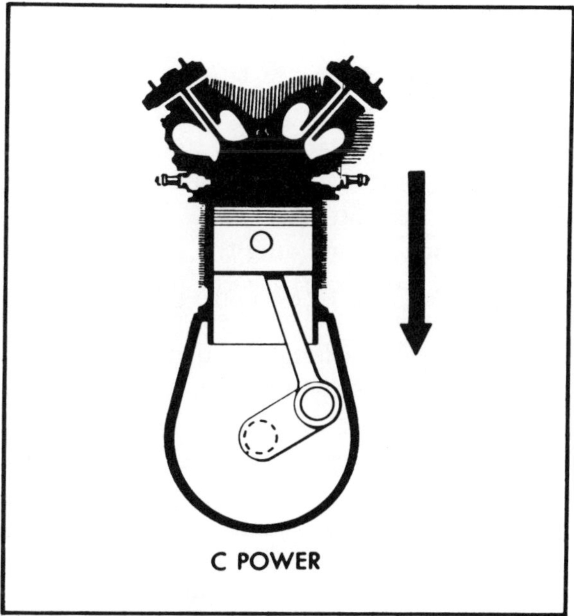

Fig. 5-32. Power stroke.

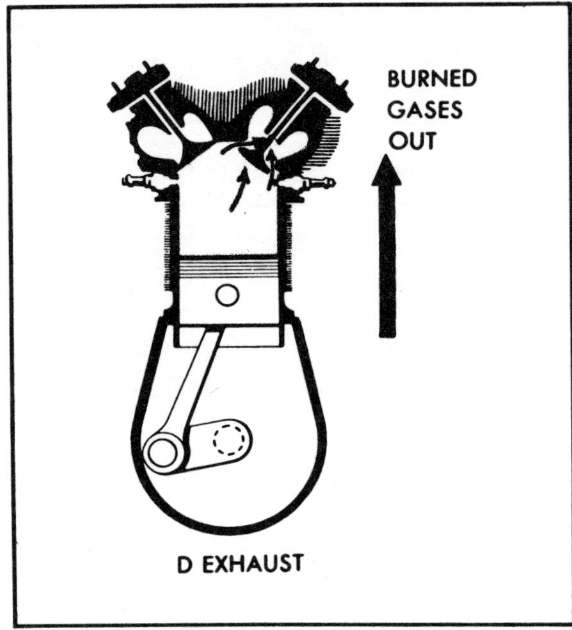

Fig. 5-33. Exhaust stroke.

downward power stroke remains open during the subsequent upward stroke and allows the burned gases to be ejected from the combustion chamber or cylinders.

Engine Cooling

The engine's internal cooling is accomplished by oil circulation and its heat is carried away by airflow. When the aircraft is on the ground, very little air makes its way over the metal fins surrounding the cylinders and overheating can occur. Overheating can also be a problem when the engine is being worked hard, such as in a long climbout. The pilot should reduce rpm and increase airspeed.

Some of the dangers involved in overheating are:

- ☐ Excessive oil consumption.
- ☐ Power loss.
- ☐ Detonation and permanent internal engine damage.

As indicated above, the oil carries away a portion of the heat and as such is very important to the engine performance. The use of the manufacturer's recommended grade of oil is important. The pilot-in-command should also make sure that the oil quantity is correct before a flight. The oil pressure should be closely monitored during start, run-up, and climbout. If you notice low oil pressure during flight, check the oil temperature gauge. If the pressure is down with normal temperatures, proceed to the nearest airport and have it checked out. If the pressure is down and the temperature is up, prepare for an emergency landing—an engine failure is probably imminent.

High oil temperatures can result from:

- ☐ Low oil supply.
- ☐ Excessively high power settings.
- ☐ Climbing too steep (restricting the airflow through the cowling), especially during hot weather.
- ☐ Low octane fuel.
- ☐ Excessively lean mixture.

Aircraft Fuel and Possible Contamination

Aircraft fuel is blended and refined to specific quality standards and engine manufacturers design their engines around these standards. The octane limitations for the engine are specified in the owner's manuals and/or placards on the plane. In cases of necessity, it is acceptable to use aircraft fuel of a *higher* octane rating. Fuel of a lower octane rating than specified by the manufacturer can cause detonation and engine failure. *Never use a lower octane fuel.*

Fuel contamination can be minimized by careful preflight action. The most common contaminator is water. Water is heavier than fuel and, given sufficient time, will precipitate out to the bottom of the fuel tank. The moisture comes from water vapor in the air space just above the fuel in the tanks. At night, when the temperature drops, the moisture in that air space will condense into liquid form and settle to the bottom of the tank. For this reason, it is wise to "top off" the tanks at the end of the day to avoid the overnight problem. A flight instructor will show you how to drain off a portion of the fuel to visually check for the presence of water and other contaminants.

A lot of the "engine failures" that seemed to plague early-day aviation were actually carb ice. The temperature drop at the carburetor venturi (item E in Fig. 5-34) can be as much as 60 degrees F. because of the evaporation of the fuel in this region plus the temperature loss associated with the lower pressures in the venturi (remember Bournelli?). If the outside air temperature is between 20 and 70 degrees F., and the relative humidity is more than 50 percent, the rapid temperature drop in the carburetor can "squeeze" water out of the intake air and carb ice can occur.

Probably the greatest accumulation of carb ice occurs on the throttle plate (item D in Fig. 5-34). Ice restricts the carburetor air passage and acts like a "choke." One must be especially cautious of this problem when the throttle is partially closed, such as during a descent. Throttle valves, the venturi, and, to some degree, the intake manifold are all targets for ice formation.

Whenever icing does occur, one of the first signs is a drop in rpm (assuming a fixed-pitch propeller). Engine roughness follows quickly

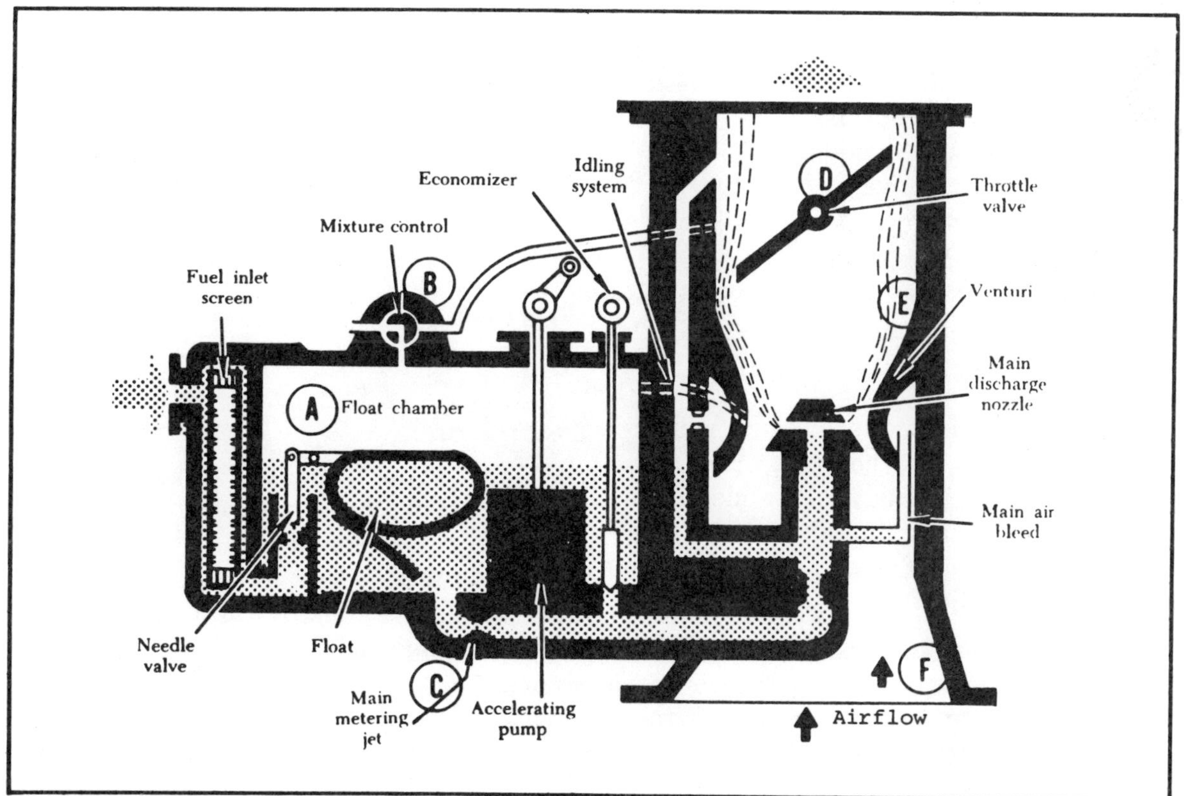

Fig. 5-34. Float-type carburetor.

and the engine can stop running in a short time. You should apply full carb heat as soon as you *suspect* that carb ice is forming. Your suspicion will be confirmed if the engine sounds *worse*! You see, the melting ice amounts to water being sucked through your engine, causing the rpm to decrease even more. As soon as the ice has been cleared, the rpm will return to near normal and stabilize. You should then turn the carb heat off and watch closely for more ice to form. Continued use of carb heat tends to increase operating temperatures and decrease engine output.

Fuel Injection

Fuel injection has several advantages over the carburetor:

- ☐ Fuel is injected at the intake valve, ensuring an equal amount of fuel to each cylinder. This means:
- ☐ Much greater efficiency.
- ☐ Better control over the mixture.
- ☐ Better starting, especially when it's quite cold.
- ☐ Because there is no fuel vaporizing at a venturi, less opportunity for icing to occur.

Fuel-injected engines sometimes will experience problems with hot start, due to vapor lock, and this gives us another pilot's rule of thumb: Never run a tank dry before switching to the next tank. (That's a good rule even *without* fuel injection.)

Fuel-Air Mixture

The mixture control (item B in Figure 5-34) is a device that controls the fuel-to-air ratio. If the mixture is too "lean" (not enough fuel for the amount of air), the engine will run rough, sometimes backfire, and lose power. A lean mixture will also have the tendency to overheat and engine damage can result from high temperatures and detonation. A lean mixture can be very detrimental during climbout or at any other time when full power is applied over an extended period.

On the other hand, a mixture that is too rich (excess fuel) can also be a problem. The extra fuel in the mixture will cause rough running, spark plug fouling, and excessive use of fuel. The mixture at idle is automatically set rich, so prolonged idling could also foul the spark plugs. But at least the overly rich mixture is not dangerous, as is the excessively lean mixture.

Aircraft engines are set so that the mixture, when full rich, gives the optimum fuel/air ratio for a standard day at sea level. The mixture is probably going to have to be leaned out as we take off and climb to altitude. This will correct the mixture for the less dense air found at higher altitudes. Here's a general rule of thumb that will get you by: Any time you will be using more than 75 percent of the engine's rated sea level horsepower, your mixture should be full rich. During most of the year at Weld County, 75 percent power is impossible.

Detonation

One of the problems that can occur due to a mixture being too lean is *detonation*. Detonation is an "explosion" instead of a "burn" of the fuel/air ratio. The explosion-created shock wave inside the cylinder can cause massive engine damage. In Fig. 5-35, the left figure shows a normal burn; the expanding gases push the piston down in a smooth stroke. Now check the explosion on the right. If the explosion occurs, it will act like a hammer. Needless to say, continued hammering can cause prob-

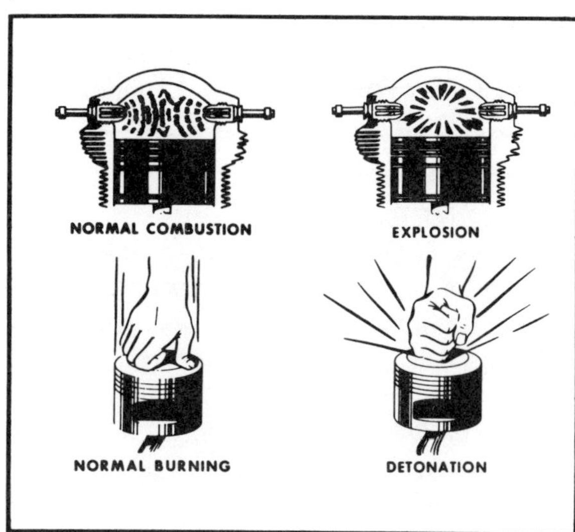

Fig. 5-35. Normal combustion and detonation.

lems, not the least of which may be engine destruction.

Detonation can be caused by engine overheating, lower octane fuel than recommended and, as just discussed, excessively lean mixtures. Rapid throttle advance, especially when the engine is overheated, could also cause detonation (plus, at any temperature, rapid power increases can "detune" the crankshaft of your engine). If you should ever suspect detonation, immediately retard the throttle, then richen the mixture and operate at reduced power loads.

Ignition and the Electrical System

The most significant fact here is that the electrical system does *not* ignite the spark plugs. The two systems are totally independent to ensure more troublefree ignition. The spark plugs are fired by magneto, like your lawn mower, and will continue to fire so long as the engine is turning over. The only exception is that if the magneto ground wire to the switch is broken, the engine will continue to run with the mag switch in the off position. Two magnetoes are used for a dual ignition system with two spark plugs per cylinder to provide improved engine performance.

The electrical system is powered by an alternator in all modern aircraft. This has an advantage over the older generators in that more power can be produced at lower rpm's. A complete electrical system failure may not be as scary as it sounds. Primarily, you will then have to operate without radio equipment (avionics).

The Constant-Speed Propeller

More advanced aircraft are equipped with a constant-speed propeller, allowing the pilot to regulate the blade angle for most efficient performance. The throttle controls power output as registered on a manifold pressure gauge and the propeller control regulates engine rpm. The pilot must be careful to avoid high manifold pressures with low rpm.

THE FAA PUBLICATIONS

Here is a quick rundown on the broad range of FAA publications intended to keep the flying public informed, safe, and legal.

Federal Aviation Regulations (FARs) are the body of federal laws passed by both the FAA and the U.S. Congress. You've already been exposed to some of these and there's more to come! Right now, notice how the "regs" are numbered by their subject areas:

FAR Part 1.	Definitions and Abbreviations.
FAR Part 61.	Certification of Pilots and Flight Instructors.
FAR Part 91.	General Operating and Flight Rules.

Airman's Information Manual (the *AIM*)

is designed to provide airmen with basic flight information and ATC procedures including information on the fundamentals required by fly in the United States National Airspace System. It also contains items of interest to pilots concerning health and medical facts, factors affecting flight safety, accident, and hazard reporting. So the AIM is full of important information, but is non-regulatory. Two major subdivisions of the AIM are:

Notice to Airmen (NOTAMs) are published to inform you of any changes of interest and/or safety in the system. The published NOTAMS can advise you of new frequencies in use, new runways, closed airports, etc. The latest publication should be brought up-to-the-minute by the FSS technician during your weather briefing. Ask for the current NOTAMs! Imagine planning a cross-country then arriving to find the airport closed!

Airport/Facility Directories contain very detailed information on airports, VORs, FSSs, etc. On the FAA test, when asked to "decode" some of this information, just refer to the complete Legend Section in the back of the test book to figure it out. Better yet, get some practice ahead of time working with the directories in planning your cross-country flights.

Some of the parts of the AIM are coming up next, but first let's check one last part of the publication system:

Advisory Circulars (AC'S): The FAA issues Advisory Circulars to provide a systematic means for the issuance of nonregulatory material of interest to the aviation public. They are issued in a numbered system of general subject matter areas to correspond with the subject areas in the Federal Aviation Regulations. For example, the Private Pilot—Airplane Written Test is published as AC 61-32C (see FAR 61, above). An article about wake turbulence would be published in the "90 series" of ACs, etc.

EXCERPTS FROM THE AIRMAN'S INFORMATION MANUAL

Wake Turbulence

540. GENERAL

a. Every airplane generates a wake while in flight. Initially, when pilots encountered this wake in flight, the disturbance was attributed to "prop wash." It is known, however, that this disturbance is caused by a pair of counter rotating vortices trailing from the wing tips. The vortices from large aircraft pose problems to encountering aircraft. For instance, the wake of these aircraft can impose rolling movements exceeding the roll control capability of some aircraft. Further, turbulence generated within the vortices can damage aircraft components and equipment if encountered at close range. The pilot must learn to envision the location of the vortex wake generated by large aircraft and adjust his flight path accordingly.

b. During ground operations, jet engine blast (thrust stream turbulence) can cause damage and upsets if encountered at close range. Exhaust velocity versus distance studies at various thrust levels have shown a need for light aircraft to maintain an adequate separation during ground operations. Below are examples of the distance requirements to avoid exhaust velocities of greater than 25 mph:

25 MPH VELOCITY	B-727	DC-8	DC-10
Takeoff Thrust	550 Ft.	700 Ft.	2100 Ft.
Breakaway Thrust	200 Ft.	400 Ft.	850 Ft.
Idle Thrust	150 Ft.	35 Ft.	350 Ft.

c. Engine exhaust velocities generated by large jet aircraft during initial takeoff roll and the drifting of the turbulence in relation to the crosswind component dictate the desirability of lighter aircraft awaiting takeoff to hold well back of the runway edge of taxiway hold line; also, the desirability of aligning the aircraft to face the possible jet engine blast movement. Additionally, in the course of running up engines and taxiing on the ground, pilots of large aircraft in particular should consider the effects of their jet blasts on other aircraft.

d. The FAA has established new standards for the location of taxiway hold lines at airports served by air carriers as follows:

e. Taxiway holding lines will be established at 100 feet from the edge of the runway, except at locations where heavy jets will be operating, the taxiway holding line markings will be established at 150 feet. (The heavy category can include some B-707 and DC-8 type aircraft.)

541. VORTEX GENERATION

Lift is generated by the creation of a pressure differential over the wing surface. The lowest pressure occurs over the upper wing surface and the highest pressure under the wing. This pressure differential triggers the rollup of the airflow aft of the wing resulting in swirling air masses trailing downstream of the wing tips. After the roll up is completed, the wake consists of two counter rotating cylindrical vortices.

542. VORTEX STRENGTH

a. The strength of the vortex is governed by the weight, speed, and shape of the wing of the generating aircraft. The vortex characteristics of any given aircraft can also be changed by extension of flaps or other wing configuring devices as well as by change in speed. However, as the basic factor is weight, the vortex strength increases proportionately. During tests, peak vortex tangential velocities were recorded at 224 feet per second, or about 133 knots. The greatest vortex strength occurs when the generating aircraft is HEAVY, CLEAN, and SLOW.

b. INDUCED ROLL

(1) In rare instances a wake encounter could cause in flight structural damage of catastrophic proportions. However, the usual hazard is associated with induced rolling movements which can exceed the rolling capability of the encountering aircraft. In flight experiments, aircraft have been intentionally flown directly up trailing vortex cores of large aircraft. It was shown that the capability of an aircraft to counteract the roll imposed by the wake vortex primarily depends on the wing span and counter responsiveness of the encountering aircraft.

(2) Counter control is usually effective and induced roll minimal in cases where the wing span and ailerons of the encountering aircraft extend beyond the rotational flow field of the vortex. It is more difficult for aircraft with short wing span (relative to the generating aircraft) to counter the imposed roll induced by vortex flow. Pilots of short span aircraft, even of the high performance type, must be especially alert to vortex encounters.

(3) The wake of large aircraft requires the respect of all pilots.

543. VORTEX BEHAVIOR

a. Trailing vortices have certain behavioral characteristics which can help a pilot visualize the wake locations and thereby take avoidance precautions.

(1) Vortices are generated from the moment aircraft leave the ground, since trailing vortices are a by-product of wing lift. Prior to takeoff or touchdown pilots should note the rotation or touchdown point of the preceding aircraft.

(2) The vortex circulation is outward, upward and around the wing tips when

viewed from either ahead or behind the aircraft. Tests with large aircraft have shown that the vortex flow field, in a plane cutting through the wake at any point downstream, covers an area about 2 wing spans in width and one wing span in depth. The vortices remain so spaced (about a wing span apart) even drifting with the wind, at altitudes greater than a wing span from the ground. In view of this, if persistent vortex turbulence is encountered, a slight change of altitude and lateral position (preferably upwind) will provide a flight path clear of the turbulence.

(3) Flight tests have shown that the vortices from large aircraft sink at a rate of about 400 to 500 feet per minute. They tend to level off at a distance about 900 feet below the flight path of the generating aircraft. Vortex strength diminishes with time and distance behind the generating aircraft. Atmospheric turbulence hastens breakup. Pilots should fly at or above the large aircraft's flight path, altering course as necessary to avoid the area behind and below the generating aircraft.

(4) When the vortices of large aircraft sink close to the ground (within about 200 feet), they tend to move laterally over the ground at a speed of about 5 knots.

b. A crosswind will decrease the lateral movement of the upwind vortex and increase the movement of the downwind vortex. Thus a light wind of 3 to 7 knots could result in the upwind vortex remaining in the touchdown zone for a period of time and hasten the drift of the downwind vortex toward another runway. Similarly, a tailwind condition can move the vortices of the preceding aircraft forward into the touchdown zone. THE LIGHT QUARTERING TAILWIND REQUIRES MAXIMUM CAUTION. Pilots should be alert to large aircraft upwind from their approach and takeoff flight paths.

544. OPERATIONS PROBLEM AREAS

a. A wake encounter is not necessarily hazardous. It can be one or more jolts with varying severity depending upon the direction of the encounter, distance from the generating aircraft, and point of vortex encounter. The probability of induced roll increases when the encountering aircraft's heading is generally aligned with the vortex trail or flight path of the generating aircraft.

b. AVOID THE AREA BELOW AND BEHIND THE GENERATING AIRCRAFT, ESPECIALLY AT LOW ALTITUDE WHERE EVEN A MOMENTARY WAKE ENCOUNTER COULD BE HAZARDOUS.

c. Pilots should be particularly alert in calm wind conditions and situations where the vortices could:

(1) Remain in the touchdown area.

(2) Drift from aircraft operating on a nearby runway.

(3) Sink into the takeoff or landing path from a crossing runway.

(4) Sink into the traffic patterns from other airport operations.

(5) Sink into the flight path of VFR flights operating at 500 feet AGL and below.

d. Pilots of all aircraft should visualize the location of the vortex trail behind large aircraft and use proper vortex avoidance procedures to achieve safe operation. It is equally important that pilots of large aircraft plan or adjust their flight paths to minimize vortex exposure to other aircraft.

545. VORTEX AVOIDANCE PROCEDURES

a. Under certain conditions, airport traffic controllers apply procedures for separating aircraft from heavy jet aircraft. The controllers will also provide VFR aircraft, with whom they are in communication and, which in the tower's opinion may be adversely affected by wake turbulence from a large aircraft, the position, altitude and direction of flight of the large aircraft followed by the phrase "CAUTION - WAKE TURBULENCE." WHETHER OR NOT A WARNING HAS BEEN GIVEN, HOWEVER, THE PILOT IS EXPECTED TO ADJUST HIS OPERATIONS AND FLIGHT PATH AS NECESSARY TO PRECLUDE SERIOUS WAKE ENCOUNTERS.

b. The following vortex avoidance procedures are recommended for the various situations:

(1) Landing behind a large aircraft-same runway: Stay at or above the large aircraft's final approach flight path-note his touchdown point-land beyond it.

(2) Landing behind a large aircraft-when parallel runway if closer than 2,500 feet: Consider possible drift to your runway. Stay at or above the large aircraft's final approach flight path-note his touchdown point.

(3) Landing behind a large aircraft-crossing runway: Cross above the large aircraft's flight path.

(4) Landing behind a departing large aircraft-same runway: Note the large aircraft's rotation point-land well prior to rotation point.

(5) Landing behind a departing large aircraft-crossing runway: Note the large aircraft's rotation point-if past the intersection-continue the approach-land prior to the intersection. If large aircraft rotates prior to the intersection, avoid flight below the large aircraft's flight path. Abandon the approach unless a landing is assured well before reaching the intersection.

(6) Departing behind a large aircraft: Note the large aircraft's rotation point-rotate prior to large aircraft's rotation point-continue climb above and stay upwind of the large aircraft's climb path until turning clear of his wake. Avoid subsequent headings which will cross below and behind a large aircraft. Be alert for any critical takeoff situation which could lead to a vortex encounter.

(7) Intersection takeoffs-same runway: Be alert to adjacent large aircraft operations particularly upwind of your runway. If intersection takeoff clearance is received, avoid subsequent heading which will cross below a large aircraft's path.

(8) Departing or landing after a large aircraft executing a low approach, missed approach or touch-and-go landing: Because vortices settle and move laterally near the ground, the vortex hazard may exist along the runway and in your flight path after a large aircraft has executed a low approach, missed approach or a touch-and-go landing, particular in light quartering wind conditions. You should assure that an interval of at least 2 minutes has elapsed before your takeoff or landing.

(9) Enroute VFR (thousand-foot altitude plus 500 feet): Avoid flight below and behind a large aircraft's path. If a large aircraft is observed above on the same track (meeting or overtaking) adjust your position laterally, preferably upwind.

550. HELICOPTERS

a. A hovering helicopter generates a downwash from its main rotor(s) similar to the prop blast of a conventional aircraft. However, in forward flight, this energy is transformed into a pair of trailing vortices similar to wingtip vortices of fixed wing aircraft. Pilots of small aircraft should avoid the vortices as well as the downwash.

551. PILOT RESPONSIBILITY

a. Government and industry groups are making concerted efforts to minimize or eliminate the hazards of trailing vortices. However, the flight disciplines necessary to assure vortex avoidance during VFR operations must be exercised by the pilot. Vortex visualization and avoidance procedures should be exercised by the pilot using the same degree of concern as in collision avoidance.

b. Wake turbulence may be encountered by aircraft in flight as well as when operating on the airport movement area. (See WAKE TURBULENCE in the Pilot/Controller glossary).

c. Pilots are reminded that in operations conducted behind all aircraft, acceptance of instructions from ATC in the following situations is an acknowledgment that the pilot will ensure safe takeoff and landing intervals and accepts the responsibility of providing his own wake turbulence separation.

(1) Traffic information.

(2) Instructions to follow an aircraft, and

(3) The acceptance of a visual approach clearance.

d. For operations conducted behind heavy aircraft, ATC will specify the word 'heavy' when this information is known. Pilots of heavy aircraft should always use the word 'heavy' in radio communications.

552. AIR TRAFFIC WAKE TURBULENCE SEPARATIONS

a. Because of the possible effects of wake turbulence, controllers are required to apply no less than specified minimum separation for aircraft operating behind a heavy jet and, in certain instances, behind large nonheavy aircraft.

(1) Separation is applied to aircraft operating directly behind a heavy jet at the same altitude or less than 1,000 feet below:

(a) Heavy jet behind heavy jet-5 miles.

(b) Small/large aircraft behind heavy jet-5 miles.

(2) Also, separation, measured at the time the preceding aircraft is over the landing threshold, is provided to small aircraft:

(a) Small aircraft landing behind heavy jet-6 miles.

(b) Small aircraft landing behind large aircraft-4 miles.

NOTE-See Aircraft Classes in Pilot/Controller Glossary.

(3) Additionally, appropriate time or distance intervals are provided to departing aircraft:

(a) Two minutes or the appropriate 4 or 5 mile radar separation when takeoff behind a heavy jet will be:

—from the same threshold

—on a crossing runway and projected flight path will cross

—from the threshold of a parallel runway when staggered ahead of that of the adjacent runway by less than 500 feet and when the runways are separated by less than 2,500 feet

NOTE-Pilots, after considering possible wake turbulence effects, may specifically request waiver of the 2-minute interval by stating, "request waiver of 2-minute interval" or a similar statement. Controllers may acknowledge this statement as pilot acceptance of responsibility for wake turbulence separation and, if traffic permits, issue takeoff clearance.

b. A 3-minute interval will be provided when a small aircraft will takeoff from an intersection after a large aircraft or heavy jet departs (same runway or a parallel runway separated by less than 2,500 feet). Controllers may not waive the time interval behind a heavy jet but may upon specific pilot request waive this 3-minute interval when the preceding aircraft is a large (nonheavy) aircraft.

Medical Facts for Pilots

FITNESS FOR FLIGHT

a. Medical Certification

(1) All pilots except those flying gliders and free air balloons must possess valid medical certificates in order to exercise the privileges of their airman certificates. The periodic medical examinations required for medical certification are conducted by designated Aviation Medical Examiners, who are physicians with a special interest in aviation safety and training in aviation medicine.

(2) The standards for medical certification are contained in FAR 67. Pilots who have a history of certain medical conditions described in these standards are mandatorily disqualified from flying. These medical conditions include a personality disorder manifested by overt acts, a psychosis, alcoholism, drug dependence, epilepsy, an unexplained disturbance of consciousness, myocardial infraction, angina pectoris and diabetes requiring medication for its control. Other medical conditions may be temporarily disqualifying, such as acute infections, anemia, and peptic ulcer. Pilots who do not meet medical standards may still be qualified under special issuance provisions or the exemption process. This may require that either additional medical information be provided or practical flight tests be conducted.

(3) Student pilots should visit an Aviation Medical Examiner as soon as possible in their flight training in order to avoid unnecessary training expenses should they not meet the medical standards. For the same reason, the student pilot who plans to enter commercial aviation should apply for the highest class of medical certificate that might be necessary in the pilot's career.

Caution: The Federal Aviation Regulations prohibit a pilot who possesses a current medical certificate from performing crewmember duties while the pilot has a known medical condition or increase of a known medical condition that would make the pilot unable to meet the standards for the medical certificate.

b. Illness

(1) Even a minor illness suffered in day-to-day living can seriously degrade performance of many piloting tasks vital to safe flight. Illness can produce fever and distracting symptoms that can impair judgment, memory, alertness, and the ability to make calculations. Although symptoms from an illness may be under adequate control with a medication, the medication itself may decrease pilot performance.

(2) The safest rule is not to fly while suffering from any illness. If this rule is considered too stringent for a particular illness, the pilot should contact an Aviation Medical Examiner for advice.

c. Medication

(1) Pilot performance can be seriously degraded by both prescribed and over-the-counter medications, as well as by the medical conditions for which they are taken. Many medications, such as tranquilizers, sedatives, strong pain relievers, and cough-suppressant preparations, have primary effects that may impair judgement, memory, alertness, coordination, vision, and the ability to make calculations. Others, such as antihistamines, blood pressure drugs, muscle relaxants, and agents to control diarrhea and motion sickness, have side effects that may impair the same critical functions. Any medication that depresses the nervous system, such as a sedative, tranquilizer or antihistamine, can make a pilot much more susceptible to hypoxia (see below).

(2) The Federal Aviation Regula-

tions prohibit pilots from performing crewmember duties while using any medication that affects the faculties in any way contrary to safety. The safest rule is not to fly as a crewmember while taking any medication, unless approved to do so by the FAA.

d. Alcohol

e. Extensive research has provided a number of facts about the hazards of alcohol consumption and flying. As little as one ounce of liquor, one bottle of beer or four ounces of wine can impair flying skills, with the alcohol consumed in these drinks being detectable in the breath and blood for at least three hours. Even after the body completely destroys a moderate amount of alcohol, a pilot can still be severely impaired for many hours by hangover. There is simply no way of increasing the destruction of alcohol or alleviating a hangover. Alcohol also renders a pilot much more susceptible to disorientation and hypoxia (see below).

(1) A consistently high alcohol related fatal aircraft accident rate serves to emphasize that alcohol and flying are a potentially lethal combination. The Federal Aviation Regulations prohibit pilots from performing crewmember duties within eight hours after drinking any alcoholic beverage or while under the influence of alcohol. However, due to the slow destruction of alcohol, a pilot may still be under the influence eight hours after drinking a moderate amount of alcohol. Therefore, an excellent rule is to allow at least 12 to 24 hours between "bottle and throttle," depending on the amount of alcoholic beverage consumed.

f. Fatigue

(1) Fatigue continues to be one of the most treacherous hazards to flight, safety, as it may not be apparent to a pilot until serious errors are made. Fatigue is best described as either acute(short-term) or chronic (long-term).

(2) A normal occurrence of everyday living, acute fatigue is the tiredness felt after long periods of physical and mental strain, including strenuous muscular effort, immobility, heavy mental workload, strong emotional pressure, monotony and lack of sleep. Consequently, coordination and alertness, so vital to safe pilot performance, can be reduced. Acute fatigue is prevented by adequate rest and sleep, as well as regular exercise and proper nutrition.

(3) Chronic fatigue occurs when there is not enough time for full recovery between episodes of acute fatigue. Performance continues to fall off, and judgment becomes impaired so that unwarranted risks may be taken. Recovery from chronic fatigue requires a prolonged period of rest.

g. Stress

(1) Stress from the pressures of everyday living can impair pilot performance, often in very subtle ways. Difficulties, particularly at work, can occupy thought processes enough to markedly decrease alertness. Distraction can so interfere with judgment that unwarranted risks are taken, such as flying into deteriorating weather conditions to keep on schedule. Stress and fatigue (see above) can be an extremely hazardous combination.

(2) Most pilots do not leave stress "on the ground." Therefore when more than usual difficulties are being experienced, a pilot should consider delaying flight until these difficulties are satisfactorily resolved.

h. Emotion

(1) Certain emotionally upsetting events, including a serious argument, death of a family member, separation or divorce, loss of job and financial catastrophe, can render a pilot unable to fly an aircraft safely. The emo-

tions of anger, depression, and anxiety from such events not only decrease alertness but also may lead to taking risks that border on self-destruction. Any pilot who experiences an emotionally upsetting event should not fly until satisfactorily recovered from it.

i. Personal Checklist

(1) Aircraft accident statistics show that pilots should be conducting preflight checklists on themselves as well as their aircraft, for pilot impairment contributes to many more accidents than failures of aircraft systems. A personal checklist that can be easily committed to memory, which includes all of the categories of pilot impairment as discussed in this section, is being distributed by the FAA in the form of a wallet-sized card.

PERSONAL CHECKLIST

I'm physically and mentally safe to fly-not being impaired by:

Illness,
Medication,
Stress,
Alcohol,
Fatigue,
Emotion.

EFFECTS OF ALTITUDE

a. Hypoxia

(1) *Hypoxia is a state of oxygen deficiency in the body sufficient to impair functions of the brain and other organs.* Hypoxia from exposure to altitude is due only to the reduced barometric pressures encountered at altitude, for the concentration of oxygen in the atmosphere remains about 21 percent from the ground out to space.

(2) Although a deterioration in night vision occurs at a cabin pressure altitude as low as 5,000 feet, other significant effects of altitude hypoxia usually do not occur in the normal healthy pilot below 12,000 feet. From 12,000 to 15,000 feet of altitude, *judgment, memory, alertness, coordination and ability to make calculations are impaired, and headache, drowsiness, dizziness and either a sense of well-being (euphoria) or belligerence occur.* The effects appear following increasingly shorter periods of exposure to increasing altitude. In fact, pilot performance can seriously deteriorate within 15 minutes at 15,000 feet.

(3) At cabin pressure altitudes above 15,000 feet, the periphery of the visual field grays out to a point where only central vision remains (tunnel vision). A blue coloration (cyanosis) of the fingernails and lips develops. The ability to take corrective and protective action is lost in 20 to 30 minutes at 18,000 feet and 5 to 12 minutes at 20,000 feet, followed soon thereafter by unconsciousness.

(4) The altitude at which significant effects of hypoxia occur can be lowered by a number of factors. Carbon monoxide inhaled in smoking or from exhaust fumes (see below), lowered hemoglobin (anemia), and certain medications can reduce the oxygen-carrying capacity of the blood to the degree that the amount of oxygen provided to body tissues will already be equivalent to the oxygen provided to the tissues when exposed to a cabin pressure altitude of several thousand feet. Small amounts of alcohol and low doses of certain drugs, such as antihistamines, tranquilizers, sedatives and analgesics can, through their depressant sections, render the brain much more susceptible to hypoxia. Extreme heat and cold fever, and anxiety increase the body's demand for oxygen, and hence its susceptibility to hypoxia.

(5) *The effects of hypoxia are usually quite difficult to recognize, especially when they occur gradually.* Since symptoms of hypoxia do not vary in an individual, the ability to

recognize hypoxia can be greatly improved by experiencing and witnessing the effects of hypoxia during an altitude chamber "flight." The FAA provides this opportunity through aviation physiology training, which is conducted at the FAA Civil Aeromedical Institute and at many military facilities across the United States. Pilots can apply for this training by contacting the Physiological Operations and Training Section, AAC-143, FAA Civil Aeromedical Institute, P.O. Box 25082, Oklahoma City, Oklahoma 73125.

(6) Hypoxia is prevented by heeding factors that reduce tolerance to altitude, by enriching the inspired air with oxygen from an appropriate oxygen system and by maintaining a comfortable, safe cabin pressure altitude. *For optimum protection, pilots are encouraged to use supplemental oxygen above 10,000 feet during the day, and above 5,000 feet at night.* The Federal Aviation Regulations require that the minimum flight crew be provided with and use supplemental oxygen after 30 minutes of exposure to cabin pressure altitudes between 12,500 and 14,000 feet, and immediately on exposure to cabin pressure altitudes above 14,000 feet. Every occupant of the aircraft must be provided with supplemental oxygen at cabin pressure altitudes above 15,000 feet.

b. Ear Block

(1) As the aircraft cabin pressure decreases during ascent, the expanding air in the middle ear pushes the eustachian tube open and, by escaping down it to the nasal passages, equalizes in pressure with the cabin pressure. But during descent, the pilot must periodically open the eustachian tube to equalize pressure. This can be accomplished by swallowing, yawning, tensing muscles in the throat or, if these do not work, by the combination of closing the mouth, pinching the nose closed and attempting to blow through the nostrils (Valsalva maneuver).

(2) Either an upper respiratory infection, such as a cold or sore throat, or a nasal allergic condition can produce enough congestion around the eustachian tube to make equalization difficult. Consequently, the difference in pressure between the middle ear and aircraft cabin can build up to a level that will hold the eustachian tube closed, making equalization difficult if not impossible. The problem is commonly referred to as an "ear block."

(3) An ear block produces severe ear pain and loss of hearing that can last from several hours to several days. Rupture of the ear drum can occur in flight or after landing. Fluid can accumulate in the middle ear and become infected.

(4) An air block is prevented by not flying with an upper respiratory infection or nasal allergic condition. Adequate protection is usually not provided by decongestant sprays or drops to reduce congestion around the eustachian tubes. Oral decongestants have side effects that can significantly impair pilot performance.

(5) If an ear block does not clear shortly after landing, a physician should be consulted.

c. Sinus Block

(1) During ascent and descent, air pressure in the sinuses equalizes with the aircraft cabin pressure through small openings that connect the sinuses to the nasal passages. Either an upper respiratory infection, such as a cold or sinusitis, or a nasal allergic condition can produce enough congestion around an opening to slow equalization and, as the difference in pressure between the sinus and cabin mounts, eventually plug the opening. This "sinus block" occurs most frequently during descent.

(2) A sinus block can occur in the frontal sinuses, located above each eyebrow, or in the maxillary sinuses, located in each upper cheek. It will usually produce excruciating pain over the sinus area. A maxillary sinus block can also make the upper teeth ache. Bloody mucus may discharge from the nasal passages.

(3) A sinus block is prevented by not flying with a respiratory infection or nasal allergic condition. Adequate protection is usually not provided by decongestant sprays or drops to reduce congestion around the sinus openings. Oral decongestants have side effects that can impair pilot performance.

(4) If a sinus block does not clear shortly after landing, a physician should be consulted.

d. Decompression Sickness After Scuba Diving

(1) A pilot or passenger who intends to fly after Scuba diving should allow the body sufficient time to rid itself of excess nitrogen absorbed during diving. If not, decompression sickness due to evolved gas can occur during exposure to low altitude and create a serious inflight emergency.

(2) The recommended waiting time before flight to cabin pressure altitudes of 8,000 feet or less is at least 2 hours after diving which has not required controlled ascent (non-decompression diving), and at least 24 hours after diving which has required controlled ascent (decompression diving). The waiting time before flight to cabin pressure altitudes above 8,000 feet should be at least 24 hours after any Scuba diving.

HYPERVENTILATION IN FLIGHT

a. Hyperventilation, or an abnormal increase in the volume of air breathed in and out of the lungs, can occur subconsciously when a stressful situation is encountered in flight. As hyperventilation "blows off" excessive carbon dioxide from the body, a pilot can experience symptoms of lightheadness, suffocation, drowsiness, tingling in the extremities, and coolness-and react to them with even greater hyperventilation. Incapacitation can eventually result from incoordination, disorientation, and painful muscle spasms. Finally, unconsciousness can occur.

b. The symptoms of hyperventilation subside within a few minutes after the rate and depth of breathing are consciously brought back under control. The buildup of carbon dioxide in the body can be hastened by controlled breathing in and out of a paper bag held over the nose and mouth.

c. Early symptoms of hyperventilation and hypoxia are similar. Moreover, hyperventilation and hypoxia can occur at the same time. Therefore, if a pilot is using an oxygen system when symptoms are experienced, the oxygen regulator should immediately be set to deliver 100 percent oxygen, and then the system checked to assure that it has been functioning effectively before giving attention to rate and depth of breathing.

CARBON MONOXIDE POISONING IN FLIGHT

a. Carbon monoxide is a colorless, odorless and tasteless gas contained in exhaust fumes. When breathed even in minute quantities over a period of time, it can significantly reduce the ability of the blood to carry oxygen. Consequently, effects of hypoxia occur (see above).

b. Most heaters in light aircraft work by air flowing over the manifold. Use of these heaters while exhaust fumes are escaping through manifold cracks and seals is responsible every year for several non-fatal and fatal aircraft accidents from carbon monoxide poisoning.

c. A pilot who detects the odor of exhaust or experiences symptoms of headache, drowsiness, or dizziness while using the heater should suspect carbon monoxide poisoning, and immediately shut off the heater and open air vents. If symptoms are severe, or continue after landing, medical treatment should be sought.

ILLUSIONS IN FLIGHT

a. Introduction

(1) Many different illusions can be experienced in flight. Some can lead to spatial disorientation. Others can lead to landing errors. Illusions rank among the most common factors cited as contributing to fatal aircraft accidents.

b. ILLUSIONS Leading to Spatial Disorientation

(1) Various complex motions and forces and certain visual scenes encountered in flight can crease illusions of motion and position. Spatial disorientation from these illusions can be prevented only by visual reference to reliable, fixed points on the ground or to flight instruments.

(2) The leans—An abrupt correction of a banked attitude, which has been entered too slowly to stimulate the motion sensing system in the inner ear, can create of illusion of banking in the opposite direction. The disoriented pilot will roll the aircraft back into its original dangerous attitude or, if level flight is maintained, will feel compelled to lean in the preceived vertical plane until this illusion subsides.

(a) Coriolis illusion—An abrupt head movement in a prolonged constant-rate turn that has ceased stimulating the motion sensing system can create the illusion of rotation or movement in an entirely different axis. The disoriented pilot will maneuver the aircraft into a dangerous attitude in an attempt to stop rotation. This most overwhelming of all illusions in flight may be prevented by not making sudden, extreme head movements, particularly while making prolonged constant-rate turns under IFR conditions.

(b) Graveyard spin—A proper recovery from a spin that has ceased stimulating the motion sensing system can create the illusion of spinning in the opposite direction. The disoriented pilot will return the aircraft to its original spin.

(c) Graveyard spiral—An abserved loss of altitude during a coordinated constant-rate turn that has ceased stimulating the motion sensing system can create the illusion of being in a descent with the wings level. The disoriented pilot will pull back on the controls, tightening the spiral and increasing the loss of altitude.

(d) Somatogravic illusion—A rapid acceleration during takeoff can create the illusion of being in a nose-up attitude. The disoriented pilot will push the aircraft into a nose-low, or dive attitude. A rapid deceleration by a quick reduction of the throttles can have the opposite effect, with the disoriented pilot pulling the aircraft into a nose-up, or stall attitude.

(e) Inversion illusion—An abrupt change from climb to straight and level flight can create the illusion of tumbling backwards. The disoriented pilot will push the aircraft abruptly into a now-low attitude, possibly intensifying this illusion.

(f) Elevator illusion—An abrupt upward vertical acceleration, usually by an updraft, can create the illusion of being in a climb. The disoriented pilot will push the aircraft into a nose-low attitude. An abrupt downward vertical acceleration, usually by a downdraft, has the opposite effect, with the disoriented pilot pulling the aircraft into a nose-up attitude.

(g) False horizon—Sloping cloud formations, an abscured horizon, a dark scene spread with ground lights and stars, and certain geometric patterns of ground light can create illusions of not being aligned correctly with the actual horizon. The disoriented pilot will place the aircraft in a dangerous attitude.

(h) Autokinesis—In the dark, a static light will appear to move about when stared at for many seconds. The disoriented pilot will lose control of the aircraft in attempting to align it with the light.

(3) Illusion Leading to Landing Errors

(a) Various surface features and atmospheric conditions encountered in landing can create illusions of incorrect height above and distance from the runway threshold. Landing errors from these illusions can be prevented by anticipating them during approaches, aerial visual inspection of unfamiliar airports before landing, using electronic glideslope or VASI systems when available, and maintaining optimum proficiency in landing procedures.

(b) Runway width illusion—a narrower than usual runway can create the illusion that the aircraft is at a higher altitude than it is actually at. The pilot who does not recognize this illusion will fly a lower approach, with the risk of striking objects along the approach path or landing short. A wider-than-usual runway can have the opposite effect,with the risk of leveling out high and landing hard or overshooting the runway.

(c) Runway and terrain slopes illusion—An upsloping runway, upsloping terrain, or both, can create the illusion that the aircraft is at a higher altitude than it is actually at. The pilot who does not recognize this illusion will fly a lower approach. A downsloping runway, downsloping approach terrain, or both, can have the opposite effect.

(d) Featureless terrain illusion—An absence of ground features, as when landing over water, darkened areas and terrain made featureless by snow, can create the illusion that the aircraft is at a higher altitude than it is actually at. The pilot who does not recognize this illusion will fly a lower approach.

(e) Atmospheric illusions—Rain on the windscreen can create the illusion of greater height, and atmospheric haze the illusion of being at a greater distance from the runway. The pilot who does not recognize these illusions will fly a lower approach. Penetration of fog can create the illusion of pitching up. The pilot who does not recognize this illusion will steepen the approach, often quite abruptly.

(f) Ground lighting illusions—Lights along a straight path, such as a road, and even lights on moving trains can be mistaken for runway and approach lights. Bright runway and approach lighting systems, especially where few lights illuminate the surrounding terrain, may create the illusion of less distance to the runway. The pilot who does not recognize this illusion will fly a higher approach.

VISION IN FLIGHT

a. Introduction

(1) Of the body senses, vision is the most important for safe flight. Major factors that determine how effectively vision can be used are the level of illumination and the technique of scanning the sky for other aircraft.

b. Vision Under Dim and Bright Illumination

(1) Under conditions of dim illumination, small print and colors on aeronautical charts and aircraft instruments become unreadable unless adequate cockpit lighting is available. Moreover,another aircraft must be much closer to be seen unless its navigation

lights are on.

(2) In darkness, vision becomes more sensitive to light, a process called dark adaptation. Although exposure to total darkness for at least 30 minutes is required for complete dark adaptation, the pilot can achieve a moderate degree of dark adaptation within 20 minutes under dim red cockpit lighting. Since red light severely distorts colors, especially on aeronautical charts, and can cause serious difficulty in focusing the eyes on objects inside the aircraft, its use is advisable only where optimum outside night vision capability is necessary. Even so, white cockpit lighting must be available when needed for map and instrument reading, especially under IFR conditions. Dark adaptation is impaired by exposure to cabin pressure altitudes above 5,000 feet, carbon monoxide inhaled in smoking and from exhaust fumes, deficiency of Vitamin A in the diet, and by prolonged exposure to bright sunlight. Since any degree of dark adaptation is lost with a few seconds of viewing a bright light, the pilot should close one eye when using a light to preserve some degree of night vision.

(3) Excessive illumination, especially from light reflected off the canopy, surfaces inside the aircraft, clouds,water,snow, and desert terrain, can produce glare,with uncomfortable squinting, watering of the eyes, and even temporary blindness. Sunglasses for protection from glare should absorb at least 85 percent of visible light (15 percent transmittance) and all colors equally (neutral transmittance), with negligible image distortion from refractive and prismatic errors.

c. Scanning for Other Aircraft

(1) Scanning the sky for other aircraft is a key factor in collision avoidance. It should be used continuously by the pilot and copilot (or right seat passenger) to cover all areas of the sky visible from the cockpit.

(2) Effective scanning is accomplished with a series of short, regularly spaced eye movements that bring successive areas of the sky into the central visual field. Each movement should not exceed 10 degrees, and each area should be observed for at least one second to enable detection. Although horizontal back-and-forth eye movements seem preferred by most pilots, each pilot should develop a scanning pattern that is most comfortable and then adhere to it to assure optimum scanning.

FEDERAL AVIATION REGULATIONS

Reproduced here are some important parts of the FARs that are covered on the test. Study the word of the law totally at first,then mentally say to yourself, elf, "just what does the law say?" In that way, you will get a better grasp of the idea behind the law. If you have questions, ask your ground or flight instructor.

Part 1—Definitions and Abbreviations

§ 1.1 General definitions.

As used in subchapters A through K of this chapter unless the context requires otherwise:

"*Administrator*" means the Federal Aviation Administrator or any person to whom he has delegated his authority in the matter concerned.

"*Aerodynamic coefficients*" means nondimensional coefficients for aerodynamic forces and moments.

"*Air carrier*" means a person who undertakes directly by lease, or other arrangement, to engage in air transportation.

"*Air commerce*" means interstate, overseas, or foreign air commerce or the transportation of mail by aircraft or any operation of navigation of aircraft within the limits of any Federal airway or any operation or navigation of aircraft when directly affects, or which may endanger safety in, interstate, overseas, or foreign air commerce.

"*Aircraft*" means a device that is used or intended to be used for flight in the air.

"*Aircraft engine*" means an engine that is used or intended to be used for propelling aircraft. It includes turbosuperchargers, appurtenances, and accessories necessary for its functioning, but does not include propellers.

"*Airframe*" means the fuselage, booms, nacelles, cowlings, fairings, airfoil surfaces (including rotors but excluding propellers and rotating airfoils of engines), and landing gear of an aircraft and their accessories and controls.

"*Airplane*" means an engine-driven fixed-wing aircraft heavier than air, that is supported in flight by the dynamic reaction of the air against its wings.

"*Airport*" means an area of land or water that issued or intended to be used for the landing and takeoff of aircraft, and includes its buildings and facilities, if any.

"*Airport traffic area*" means, unless otherwise specifically designated in Part 93, that airspace within a horizontal radius of 5 statute miles form the geographical center of any airport at which a control tower is operating, extending from the surface up to, but not including, an altitude of 3,000 feet above the elevation of the airport.

"*Airship*" means an engine-driven lighter-than-air aircraft that can be steered.

"*Air traffic*" means aircraft operating in the air or on an airport surface, exclusive of loading ramps and parking areas.

"*Air traffic clearance*" mens an authorization by air traffic control, for the purpose of preventing collision between known aircraft, for an aircraft to proceed under specified traffic conditions within controlled airspace.

"*Air traffic control*" means a service operated by appropriate authority to promote the safe, orderly, and expeditious flow of air traffic.

"*Approved*", unless used with reference to another person, means approved by the Administrator.

"*Balloon*" means a lighter-than-air aircraft that is not engine driven.

"*Brake horsepower*" means the power delivered at the propeller shaft (main drive or main output) of an aircraft engine.

"*Calibrated airspeed*" means indicated airspeed of an aircraft, corrected for position and instrument error. Calibrated airspeed is equal to true airspeed in standard atmosphere at sea level.

"*Category*"-

(1) As used with respect to the certification, ratings, privileges, and limitations of airmen, means a broad classification of aircraft. Examples include: airplane; rotorcraft; glider; and lighter-than-air; and

(2) As used with respect to the certification of aircraft, means a grouping of aircraft based upon intended use of operating limitations. Examples include: transport; normal; utility; acrobatic; limited; restricted; and provisional.

"*Ceiling*" means the height above the earth's surface of the lowest layer of clouds or obscuring phenomena that is reported as "broken", "overcase", or "obscuration", and not classified as "thin" or "partial".

"*Class*"—

(1) As used with respect to the certification, ratings, privileges, and limitations of airmen, means a classification of aircraft within a category having similar operating characteristics. Examples in-

clude: single engine; multiengine; land; water; gyroplane; helicopter; airship; and free balloon; and

(2) As used with respect to the certification of aircraft, means a broad grouping of aircraft having similar characteristics of propulsion, flight, or landing. Examples include: airplane; rotorcraft; glider; balloon; landplane; and seaplane.

"*Controlled airspace*" means airspace designated as a continental control area, control area, control zone, terminal control area, or transition area, within which some or all aircraft may be subject to air traffic control.

"*Extended over-water operation*" means—with respect to aircraft other than helicopters, and operation over water at a horizontal distance of more than 50 nautical miles form the nearest shoreline.

"*Flap extended speed*" means the highest speed permissible with wing flaps in a prescribed extended position.

"*Flight crewmember*" means a pilot, flight engineer, or flight navigator assigned to duty in an aircraft during flight time.

"*Flight level*" means a level of constant atmospheric pressure related to a reference datum of 29.92 inches of mercury. Each is stated in three digits that represent hundreds of feet. For example, flight level 250 represents a barometric altimeter indication of 25,000 feet; flight level 255, an indication of 25,500 feet.

"*Flight plan*" means specified information, relating to the intended flight of an aircraft, that is filed orally or in writing with air traffic control.

"*Flight time*" means the time from the moment the aircraft first moves under its own power for the purpose of flight until the moment it comes to rest at the next point of landing. ("Block-to-block" time.)

"*Flight visibility*" means the average forward horizontal distance, form the cockpit of an aircraft in flight, at which prominent unlighted objects may be seen and identified by day and prominent lighted objects may be seen and identified by night.

"*Helicopter*" means a rotorcraft that, for its horizontal motion, depends principally on its engine-driven rotors.

"*Indicated airspeed*" means the speed of an aircraft as shown on its pitot static airspeed indicator calibrated to reflect standard atmosphere adiabatic compressible flow at sea level uncorrected for airspeed system errors.

"*Instrument*" means a device using an internal mechanism to show visually or aurally the attitude, altitude, or operation of an aircraft or aircraft part. It includes electronic devices for automatically controlling an aircraft in flight.

"*Landing gear extended speed*" means the maximum speed at which an aircraft can be safely flown with the landing gear extended.

"*Landing gear operating speed*" means the maximum speed at which the landing gear can be safely extended or retracted.

"*Large aircraft*" means aircraft of more than 12,500 pounds, maximum certificated takeoff weight.

"*Lighter-than-air aircraft*" means aircraft that can rise and remain suspended by using contained gas weighing less than the air that is displaced by the gas.

"*Load factor*" means the ratio of a specified load to the total weight of the aircraft. The specified load is expressed in terms of any of the following: aerodynamic forces, inertia forces, or ground or water reactions.

"*Night*" means the time between the end of evening civil twilight and the beginning of morning civil twilight, as published in the

American Air Almanac, converted to local time.

"*Over-the-top—*" means above the layer of clouds or other obscuring phenomena forming the ceiling.

"*Parachute*" means a device used or intended to be used to retard the fall of a body or object through the air.

"*Person*" means an individual, firm, partnership, corporation, company, association, jointstock association, or governmental entity. It includes a trustee, receiver, assignee, or similar representative of any of them.

"*Pilotage*" means navigation by visual reference to landmarks.

"*Pilot in command*" means the pilot responsible for the operation and safety of an aircraft during flight time.

"*Positive control*" means control of all air traffic, within designated airspace, by air traffic control.

"*Preventive maintenance*" means simple or minor preservation operations and the replacement of small standard parts not involving complex assembly operations.

"*Prohibited area*" means designated airspace within which the flight of aircraft is prohibited.

"*Propeller*" means a device for propelling an aircraft that has blades on an engine-driven shaft and that, when rotated, produces by its action on the air, a thrust approximately perpendicular to its plane of rotation. It includes control components normally supplied by its manufacturer, but does not include main and auxiliary rotors or rotating airfoils of engines.

"*Rotorcraft*" means a heavier-than-air aircraft that depends principally for its support in flight on the lift generated by one or more rotors.

"*Rating*" means a statement that, as a part of a certificate, sets forth special conditions, privileges, or limitations.

"*Restricted area*" means airspace designated under Part 73 of this chapter within which the flight of aircraft, while not wholly prohibited, is subject to restriction.

"*Standard atmosphere*" means the atmosphere defined in *U.S. Standard Atmosphere, 1962* (Geopotential altitude tables).

"*Takeoff power*"—

(1) With respect to reciprocating engines, means the brake horsepower that is developed under standard sea level conditions, and under the maximum conditions of crankshaft rotational speed and engine manifold pressure approved for the normal takeoff, and limited in continuous use to the period of time shown in the approved engine specification;

"*Time in service*", with respect to maintenance time records, means the time from the moment an aircraft leaves the surface of the earth until it touches it at the next point of landing.

"*Traffic pattern*" means the traffic flow that is prescribed for aircraft landing at, taxiing on, or taking off from, an airport.

"*True airspeed*" means the airspeed of an aircraft relative to undisturbed air. True airspeed is equal to equivalent airspeed multiplied by $(po/p)^{1/2}$.

"*Type*"—

(1) As used with respect to the certification, ratings, privileges, and limitations of airmen, means a specific make and basic model of aircraft, including modifications thereto that do not change its handling or flight characteristics. Examples include: DC-7, 1049, and F-27; and

(2) As used with respect to the certification of aircraft, means those aircraft

which are similar in design. Examples include: DC-7 and DC-7C; 1049G and 1049Gand 1049H; and F-27 and F-27F.

(3) As used with respect to the certification of aircraft engines means those engines which are similar in design. For example, JT8D and JT8D-7 are engines of the same type, and JT9D-3A and JT9D-7 are engines of the same type.

"*United States*", in a geographical sense, means (1) the States, the District or Columbia, Puerto Rico, and the possessions, including the territorial waters, and (2) the airspace of those areas.

"*VFR over-the-top*", with respect to the operation of aircraft, means the operation of an aircraft over-the-top under VFR when it is not being operated on an IFR flight plan.

Part 61—Certification: Pilots and Flight Instructors

Subpart A—General

§ 61.3 Requirements for certificates, rating, and authorizations.

(a) *Pilot certificate.* No person may act as pilot in command or in any other capacity as required pilot flight crewmember of a civil aircraft of United States registry unless he has in his personal possession a current pilot certificate issued to him under this Part. However, when the aircraft is operated within a foreign country a current pilot license issued by the country in which the aircraft is operated may be used.

(b) *Pilot certificate: foreign aircraft.* No person may, within the United States, act as pilot in command or in any other capacity as a required pilot flight crewmember of a civil aircraft of foreign registry unless he has in his personal possession a current pilot certificate issued to him under this Part, or a pilot license issued to him or validated for him by the country in which the aircraft is registered.

(c) *Medical certificate.* Except for free balloon pilots piloting balloons and glider pilots piloting gliders, no person may act as pilot in command or in any other capacity as a required pilot flight crewmember of an aircraft under a certificate issued to him under this Part, unless he has in his personal possession an appropriate current medical certificate issued under Part 67 of this chapter. However, when the aircraft is operated within a foreign country with a current pilot license issued by that country, evidence of current medical qualification for that license, issued by that, may be used. In the case of a pilot license under § 61.75, evidence of current medical qualification accepted for the issue of that license is issued in place of a medical certificate.

(h) *Inspection of certificate.* Each person who holds a pilot certificate, flight instructor certificate, medical certificate, authorization, or license required by this Part shall present it for inspection upon the request of the Administrator, an authorized representative of the National Transportation Safety Board, or any Federal, State, or local law enforcement officer.

§ 61.19 Duration of pilot and flight instructor certificates.

(a) *General.* The holder of a certificate with an expiration date may not, after that date, exercise the privileges of that certificate.

(b) *Student pilot certificate.* A student pilot certificate expires at the end of the 24th month after the month in which it is issued.

(c) *Other pilot certificates.* Any pilot certificate (other than a student pilot certificate) issued under this Part is issued without a specific expiration date. However, the holder of a pilot certificate issued on the basis of a foreign pilot license may exercise the privileges of that certificate only while the foreign pilot license on which that certificate is based is effective.

(d) *Flight instructor certificate.* A flight instructor certificate-

(1) Is effective only while the holder has a current pilot certificate and a medical certificate appropriate to the pilot privileges being exercised; and

(2) Expires at the end of the 24th month after the month in which it was last issued or renewed.

(e) *Surrender, suspension, or revocation.* Any pilot certificate or flight instructor certificate issued under this Part ceases to be effective if it is surrendered, suspended, or revoked.

(f) *Return of certificate.* The holder of any certificate issued under this Part that is suspended or revoked shall, upon the Administrator's request, return it to the Administrator.

§ 61.23 Duration of medical certificates.

(a) A first-class medical certificate expires at the end of the last day of—

(1) The sixth month after the month of the date of examination shown on the certificate, for operations requiring an airline transport pilot certificate; and

(2) The 12th month after the month of the date of examination shown on the certificate, for operations requiring only a commercial pilot certificate; and

(3) The 24th month after the month of the date of examination shown on the certificate, for operations requiring only a private or student pilot certificate.

(b) A second-class medical certificate expires at the end of the last day of—

[(1) The 12th month after the month of the date of examination shown on the certificate, for operations requiring a commercial pilot certificate, or an air traffic control tower operator certificate; and]

(2) The 24th month after the month of the date of examination shown on the certificate, for operations requiring only a private or student pilot certificate.

(c) A third-class medical certificate expires at the end of the last day of the 24th month after the month of the date of examination shown on the certificate, for operations requiring a private or student pilot certificate.

§ 61.31 General limitations.

(c)*Category and class rating: carrying another person or operating for compensation or hire.* Unless he holds a category and class rating for that aircraft, a person may not act as pilot in command of an aircraft that is carrying another person or is operated for compensation or hire. In addition, he may not act as pilot in command of that aircraft for compensation or hire.

(d) *Category and class rating: other operations.* No person may act as pilot in command of an aircraft in solo flight in operations not subject to paragraph (c) of this selection, unless he meets at least one of the following:

(1) He holds a category and class rating appropriate to that aircraft.

(2) He has received flight instruction in the pilot operations required by this Part, appropriate to the category and class of aircraft for first solo, given to him by a certificated flight instructor who fund him competent to solo that category and class of aircraft and has so endorsed his pilot logbook.

(3) He has soloed and logged pilot-in-command time in that category and class of aircraft before November 1, 1973.

(e) *High performance airplanes.* A person holding a private or commercial pilot certificate may not act as pilot in command of an airplane that has more than 200 horsepower, or that has a retractable landing gear, flaps, and a controllable propeller, unless he has received flight instruction from an authorized flight instructor who has certified in his logbook that he is competent to pilot an airplane that has more than 200 horsepower, or that has a retractable landing gear, flaps, and a controllable propeller, as the case may be. However, this instruction is not required if he has logged flight time as pilot in command in high performance airplanes before November 1, 1973.

§ 61.35 Written test: prerequisites and passing grades.

(a) An applicant for a written test must—

(1) Show that he has satisfactorily completed the ground instruction or home study course required by this Part for the certificate or rating sought;

(2) Present as person identification an airman certificate, driver's license, or other official document; and

(3) Present a birth certificate or other official document showing that he meets the age requirement prescribed in this Part for the certificate sought not later than 2 years from the date of application for the test.

(b) The minimum passing grade is specified by the Administrator on each written test sheet or booklet furnished to the applicant.

§ 61.39 Prerequisites for flight tests.

(a) To be eligible for a flight test for a certificate, or an aircraft or instrument rating issued under this Part, the applicant must—

(1) Have passed any required written test since the beginning of the 24th month before the month in which he takes the flight test;

(2) Have the applicable instruction and aeronautical experience prescribed in this Part;

(3) Hold a current medical certificate appropriate to the certificate he seeks or, in the case of a rating to be added to his pilot certificate, at least a third-class medical certificate issued since the beginning of the 24th month before the month in which he takes the flight test.

§ 61.57 Recent flight experience: pilot in command.

(a) *Flight review.* After November 1,1974, no person may act as pilot in command of an aircraft unless, within the preceding 24 months, he has—

(1) Accomplished a flight review given to him, in an aircraft for which he is rated, by an appropriately certificated instructor or other person designated by the Administrator; and

(2) Had his log book endorsed by the person who gave him the review certifying that he has satisfactorily accomplished the flight review.

However, a person who has, within the preceding 24 months, satisfactorily completed a pilot proficiency check conducted by the FAA, an approved pilot check airman or a U.S. Armed Force for a pilot certificate, rating or operating privilege, need not accomplish the flight review required by this section.

(b) *Meaning of flight review.* As used in this section, a flight review consists of—

(1) A review of the current general operating and flight rules of Part 91 of this chapter; and

(2) A review of those maneuvers and procedures which in the discretion of the person giving the review are necessary for the pilot to demonstrate that he can safely exercise the privileges

of his pilot certificate.

(c) *General experience.* No person may act as pilot in command of an aircraft carrying passengers, nor of an aircraft carrying passengers, nor of an aircraft certificated for more than one required pilot flight crewmember, unless within the preceding 90 days, he has made three takeoffs and three landings as the sold manipulator of the flight controls in an aircraft of the same category and class and, if a type rating is required, of the same type. If the aircraft is a tailwheel airplane, the landings must have been made to a full stop in a tailwheel airplane. For the purpose of meeting the requirements of the paragraph a person may act as pilot-in-command of a flight under day VFR or day IFR if no persons or property other than as necessary for his compliance thereunder, are carried. This paragraph does not apply to operations requiring an airline transport pilot certificate, or to operations conducted under Part 135 of this chapter.

(d) *Night experience.* No person may act as pilot in command of an aircraft carrying passengers during the period beginning 1 hour after sunset and ending 1 hour before sunrise (as published in the American Air Almanac) unless, within the preceding 90 days, he has made at least three takeoffs and three landings to a full stop during that period in the category and class of aircraft to be used. This paragraph does not apply to operations requiring an airline transport pilot certificate.

Subpart D—Private Pilots

§ 61.103 Eligibility requirements: general.

To be eligible for a private pilot certificate, a person must—

(a) Be at least 17 years of age, except that a private pilot certificate with a free balloon or a glider rating only may be issued to a qualified applicant who is at least 16 years of age;

(b) Be able to read, speak, and understand the English language, or have such operating limitations placed on his pilot certificate as are necessary for the safe operation of aircraft, to be removed when he shows that he can read, speak, and understand the English language;

(c) Hold at least a current third-class medical certificate issued under Part 67 of this chapter, or, in the case of a glider or free balloon rating, certify that he has no known medical defect that makes him unable to pilot a glider or free balloon, as appropriate;

(d) Pass a written test on the subject areas on which instruction or home study is required by § 61.105;

(e) Pass an oral and flight test on procedures and maneuvers selected by an FAA inspector or examiner to determine the applicant's competency in the flight operations on which instruction is required by the flight proficiency provisions of § 61.07; and

(f) Comply with the section of this Part that apply to the rating he seeks.

§ 61.118 Private pilot privileges and limitations: pilot in command.

Except as provided in paragraphs (a) through (d) of this section, a private pilot may not act as pilot in command of an aircraft that is carrying passengers or property for compensation or hire; nor may he, for compensation or hire, act as pilot in command of an aircraft.

(a) A private pilot may, for compensation or hire, act as pilot in command of an aircraft in connection with any business or employment if the flight is only incidental to that business or employment and the aircraft does not carry passengers or property of compensation or hire.

(b) A private pilot may share the operating expenses of a flight with his passengers.

(c) A private pilot who is an aircraft salesman and who has at least 200 hours of logged flight time may demonstrate an aircraft in flight to a prospective buyer.

(d) A private pilot may act as pilot in command of an aircraft used in a passenger carrying airlift sponsored by a charitable organization, and for which the passengers make a donation to the organization, if—

(1) The sponsor of the airlift notifies the FAA General Aviation District Office having jurisdiction over the area concerned, at least 7 days before the fight, and furnishes any essential information that the office requests;

(2) The flight is conducted form a public airport adequate for the aircraft used, or from another airport that has been approved for the operation by an FAA inspector;

(3) He has logged at least 200 hours of flight time;

(4) No acrobatic or formation flights are conducted;

(5) Each aircraft used is certificated in the standard category and complies with the 100-hour inspection requirement of § 91.169 of this chapter; and

(6) The flight is made under VFR during the day.

For the purpose of paragraph (d) of this section, a "charitable organization" means an organization listed in Publication No. 78 of the Department of the Treasury called the "Cumulative List of Organization described in section 170(c) of the Internal Revenue Code of 1954," as amended from time to time by published supplemental listed.

Part 71—Designation of Federal Airways, Area Low Routes, Controlled Airspace, and Reporting Points

§ 71.9 Continental control area.

The continental control area consists of the airspace of the 48 contiguous States, the District of Columbia and Alaska, excluding the Alaska peninsula west of Long. 160°00′00″W., at and above 14,500 feet MSL, but does not include—

(a) The airspace less than 1,500 feet above the surface of the earth; or

[(b) Prohibited and restricted areas, other than restricted areas listed in Subpart D of this Part.]

§ 71.11 Control zones.

The control zones listed in Subpart F of this Part consist of controlled airspace which extends upward from the surface of the earth and terminates at the base of the continental control area. Control zones that do not underline the continental control area have no limit. A control zone may include one or more airports and is normally a circular area with a radius of 5 miles and any extensions necessary to include instrument approach and departure paths.

§ 71.12 Terminal control areas.

The terminal control areas listed in Subpart K of this Part consist of controlled airspace extending upward from the surface or higher to specified altitudes, within which all aircraft are subject to operating rules, pilot rules, or equipment rules specified in Part 91 of this chapter. Each such location is designated as a Group I, Group Ii, or Group III Terminal Control Area, and includes at least one primary airport around which the terminal control area is located.

§ 71.13 Transition areas.

The transition areas listed in Subpart G of this Part consist of controlled airspace extending upward form 700 feet or more above the surface of the earth when designated in conjunction with an airport for which an approved instrument approach procedure has been prescribed; or from 1,200 feet or more above the surface of the earth when designated in conjunction with airway route structures or segments. Unless otherwise specified, transition areas terminate at the base of the overlying controlled airspace.

Part 91–General Operating and Flight Rules

Subpart A—General

§ 91.3 Responsibility and authority of the pilot in command.

(a) The pilot in command of an aircraft is directly responsible for, and is the final authority as to, the operation of that aircraft.

(b) In an emergency requiring immediate action, the pilot in command may deviate from any rule of this subpart or of subpart B to the extent required to meet that emergency.

(c) Each pilot in command who deviates from a rule under paragraph 9b) of this section shall, upon the request of the administrator, send a written report of that deviation to the Administrator.

§ 91.5 Preflight action.

Each pilot in command shall, before beginning a flight, familiarize himself with all available information concerning that flight. this information must include:

(a) For a flight under IFR or a flight not in the vicinity of an airport, weather reports and forecasts, fuel requirements, alternatives available if the planned flight cannot be completed, and any known traffic delays of which he has been advised by ATC.

(b) For any flight, runway lengths at airports of intended use, and the following takeoff and landing distance information:

(1) For civil aircraft for which an approved airplane or rotorcraft flight manual containing takeoff and landing distance data is required, the takeoff and landing distance data contained therein; and

(2) For civil aircraft other than those specified in subparagraph (1) of this paragraph, other reliable information appropriate to the aircraft, relating to aircraft performance under expected values of airport elevation and runway slope, aircraft gross weight, and wind and temperature.

§ 91.7 Flight crewmembers at stations.

(a) During takeoff and landing, and while enroute, each required flight crewmember shall—

(1) Be at his station unless his absence is necessary in the performance of his duties in connection with the operation of the aircraft or in connection with his physiological needs; and

(2) Keep his seat belt fastened while at his station.

(b) After July 18, 1978, each required flight crewmember of a U.S. registered civil airplane shall, during takeoff and landing, keep the shoulder harness fastened while at his station. This paragraph does not apply if—

(1) The seat at the crewmember's station is not equipped with a shoulder harness; or

(2) The crewmember would be unable to perform his required duties with the shoulder harness fastened.

§ 91.11 Liquor and drugs.

(a) No person may act as a crewmember of a civil aircraft—

(1) Within 8 hours after the consumption of any alcoholic beverage;

(2) While under the influence of alcohol; or

(3) While using any drug that effects his faculties in any way contrary to safety.

(b) Except in an emergency, no pilot of a civil aircraft may allow a person who is obviously under the influence of intoxicating liquors or drugs (except a medical patient under proper care) to be carried in that aircraft.

§ 91.13 Dropping objects.

No pilot in command of a civil aircraft may allow any object to be dropped from that aircraft in flight that creates a hazard to persons or property. However, this section does not prohibit the dropping of any object if reasonable precautions are taken to avoid injury or damage to persons or property.

§ 91.14 Use of safety belts.

[(a) Unless otherwise authorized by the Administrator—

[(1) No pilot may take off a U.S. registered civil aircraft (except a free balloon that incorporates a basket or gondola and an airship) unless the pilot in command of that aircraft ensures that each person on board is briefed on how to fasten and unfasten that person's safety belt.]

[(2)] No pilot may take off or land a U.S. registered civil aircraft (except free balloons that incorporate baskets or gondolas and airships) unless the pilot in command of that aircraft ensures that each person on board has been notified to fasten his safety belt.

[(3)] During the takeoff and landing of U.S. registered civil aircraft (except free balloons that incorporate baskets or gondolas and airships) each person on board that aircraft must occupy a seat or berth with a safety belt properly secured about him. However, a person who has not reached his second birthday may be held by an adult who is occupying a seat or berth, and a person on board for the purpose of engaging in sport parachuting may use the floor of the aircraft as a seat.

(b) This section does not apply to operations conducted under Parts 121, 123, or 127 of this chapter. Subparagraph [(a) (3)] of this section does not apply to persons subject to § 91.7.

§ 91.22 Fuel requirements for flight under VFR.

(a) No person may begin a flight in an airplane under VFR unless (considering wind and forecase weather conditions) there is enough fuel to fly to the first point of intended landing and, assuming normal cruising speed—

(1) During the day, to fly after that for at least 30 minutes; or

(2) At night, to fly after that for at least 45 minutes.

(b) No person may begin a flight in a rotorcraft under VFR unless (considering wind and forecast weather conditions) there is enough fuel to fly to the first point of intended landing and, assuming normal cruising speed to fly after that for at least 20 minutes.

§ 91.24 [ATC transponder and altitude reporting equipment and use.]

(a) *All airspace: U.S. registered civil aircraft.* For operation not conducted under Parts 121, 123, 127, or 135 of this chapter, ATC transponder equipment installed after January 1, 1974, in U.S. registered civil aircraft not previously equipped with an ATC transponder, and all ATC transponder equipment used in U.S. registered civil aircraft July 1, 1975, must meet the performance and environmental requirements of any class of TSO-C74b or any class of TS)-C74c as appropriate, except that the Administrator may approve the use of TSO-C74 or TSO-C74a equipment after July 1, 1975, if the applicant submits data showing that such equipment meets the minimum performance standards of the appropriate class of TSO-C74c and environmental conditions of the

TSO under which it was manufactured.

(b) *Controlled airspace: all aircraft.* Except for persons operating helicopters in terminal control areas at or below 1,000 feet AGL under the terms of a letter of agreement, and except for person operating gliders above 12,500 feet MSL but below the floor of the positive control area, [no person may operate an aircraft n the controlled airspace prescribed in subparagraphs (b)(1) through (b)(4) of this paragraph], unless that aircraft is equipped with an operable coded radar beacon transponder having a Mode 3/A 4096 code capability, replying to Mode 3/A interrogation with the code specified by ATC, and is equipped with automatic pressure altitude reporting equipment having a Mode C capability that automatically replies to Mode C interrogations by transmitting pressure altitude information in 100-foot increments. This requirement applies—

(1) In Group I Terminal Control Areas governed by § 91.90(a);

(2) In Group II Terminal Control Areas governed by § 91.90(b), except as provided therein;

(3) In Group III Terminal Control Areas governed by § 91.90(c), except as provided therein; and

(4) In all controlled airspace of the 48 contiguous States and the District of Columbia, above 12,500 feet MSL, excluding the airspace at and below 2,500 feet AGL.

(c) *ATC authorized deviations.* ATC may authorize deviations from paragraph (b) of this section—

(1) Immediately, to allow an aircraft with an inoperative transponder to continue to the airport of ultimate destination, including any intermediate stops, or to proceed to a place where suitable repairs can be made, or both;

(2) Immediately, for operations of aircraft with an operating transponder but without operating transponder but without operating automatic pressure altitude reporting equipment having a Mode C capability; and

(3) On a continuing basis, or for individual flights, for operations of aircraft without a transponder, in which case the request for a deviation must be submitted to the ATC facility having jurisdiction over the airspace concerned at least four hours before the proposed operation.

§ 91.27 Civil aircraft: certifications required.

(a) Except as provided in § 91.28, no person may operate a civil aircraft unless it has within it the following:

(1) An appropriate and current airworthiness certificate. Each U.S. airworthiness certificate used to comply with this subparagraph (except a special flight permit, a copy of the applicable operations specifications issued under § 21.197(c) of this chapter, appropriate sections of the air carrier manual required by Parts 121 and 127 of this chapter containing that portion of the operations specifications issued under § 21.1987(c), or an authorization under § 91.45), must have on if the registration number assigned to the aircraft under Part 47 of this chapter. However, the airworthiness certificate need not have on it an assigned special identification number before 10 days after that number is first affixed to the aircraft. A revised airworthiness certificate having on it an assigned special identification number, that has been affixed to an aircraft, may only be obtained upon application to an FAA Flight Standards District Office.

(2) A registration certificate issued to its owner.

(b) No person may operate a civil aircraft unless the airworthiness certificate required by paragraph (a) of this section or a special flight authorization issued under § 21.28 is displayed at the cabin or cockpit entrance so that it is legible to passengers or crew.

§ 91.29 Civil aircraft airworthiness.

(a) No person may operate a civil aircraft unless it is in an airworthy condition.

(b) The pilot in command of a civil aircraft is responsible for determining whether that aircraft is in condition for safe flight. He shall discontinue the flight when unairworthy mechanical or structural conditions occur.

§ 91.31 Civil aircraft operating limitations and marking requirements.

(a) Except as provided in paragraph (d) of this section, no person may operate a civil aircraft without compliance with the operating limitations for that aircraft prescribed by the certificating authority of the country of registry.

(b) No person may operate a U.S. registered civil aircraft—

(1) For which an Airplane or Rotorcraft Flight Manual is required by § 21.5 unless there is available in the aircraft a current approved Airplane or Rotorcraft Flight Manual or the manual provided for in § 121.141(b); and

(2) For which an Airplane or Rotorcraft Flight Manual is not required by § 21.5, unless there is available in the aircraft a current approved Airplane or Rotorcraft Flight Manual, approved manual material, markings, and placards, or any combination thereof.

(c) No person may operate a U.S. registered civil aircraft unless that aircraft is identified in accordance with Part 45 of this chapter.

(d) Any person taking off or landing a helicopter certificated Under Part 29 of this chapter at a heliport constructed over water may make such momentary flight as is necessary for takeoff or landing through the prohibited range of the limiting height-speed envelope established for that helicopter if that flight through the prohibited range takes place over water on which a safe ditching can be accomplished, and if the helicopter is amphibious or is equipped with floats or other emergency flotation gear adequate to accomplish a safe emergency ditching on open water.

(e) The Airplane or Rotorcraft Flight Manual, or manual material, markings and placards required by paragraph (b) of this section must contain each operating limitation prescribed for that aircraft by the Administrator, including the following:

[(1) Powerplant (e.g., r.p.m., manifold pressure, gas temperature, etc.).

[(2) Airspeeds (e.g., normal operating speed, flaps extended speed, etc.).

[(3) Aircraft weight, center of gravity, and weight distribution, including the composition of the useful load in those combinations and ranges intended to ensure that the weight and center of gravity position will remain within approved limits (e.g., combinations and ranges of crew, oil, fuel, and baggage).

[(4) Minimum flight crew.

[(5) Kinds of operation.

[(6) Maximum operating altitude.

[(7) Maneuvering flight load factors.

[(8) Rotor speed (for rotorcraft).

[(9) Limiting height-speed envelope

(for rotorcraft)]

§ 91.32 Supplemental oxygen.

(a) *General.* No person may operate a civil aircraft of U.S. registry—

(1) At cabin pressure altitudes above 12,500 feet (MSL) up to and including 14,000 feet (MSL), unless the required minimum flight crew is provided with and uses supplemental oxygen for that part of the flight at those altitudes that is of more than 30 minutes duration;

(2) At cabin pressure altitudes above 14,000 feet (MSL), unless the required minimum flight crew is provided with and uses supplemental oxygen during the entire flight time at those altitudes; and

(3) At cabin pressure altitudes above 15,000 feet (MSL), unless each occupant of the aircraft is provided with supplemental oxygen.

§ 91.52 Emergency Locator Transmitters.

(a) Except as provided in paragraphs (e) and (f) of this section, no person may operate a U.S. registered civil airplane unless it meets the applicable requirements of paragraphs (b), (c), and (d) of this section.

(b) To comply with paragraph (a) of this section, each U.S. registered civil airplane must be equipped as follows:

(1) For operations governed by the supplemental air carrier and commercial operator rules of Part 121 of this chapter, or the air travel club rules of Part 123 of this chapter, there must be attached to the airplane an automatic type emergency locator transmitter that is an operable condition and meets the applicable requirements of [TSO-C91];

(2) For charter flights governed by the domestic and flag air carrier rules of Part 121 of this chapter, there must be attached to the airplane an automatic type emergency locator transmitter that is in operable condition and meets the applicable requirements of [TSO-C91];

(3) For operations governed by Part 135 of this chapter, there must be attached to the airplane an automatic type emergency locator transmitter that is in operable condition and meets the applicable requirements of [TSO-C91]; and

(4) For operations other than those specified in subparagraphs (1), (2), and (3) of this paragraph, there must be attached to the airplane a personnel type or an automatic type emergency locator transmitter that is in operable condition and meets the applicable requirements of [TSO-C91].

(c) Each emergency locator transmitter required by paragraphs (a) and (b) of this section must be attached to the airplane in such a manner that the probability of damage to the transmitter, in the event of crash impact, is minimized. Fixed and deployable automatic type transmitters must be attached to the airplane as far aft as practicable.

(d) Batteries used in the emergency locator transmitters required by paragraphs (a) and (b) of this section must be replaced (or recharged, if the battery is rechargeable)—

(1) When the transmitter has been in use for more than one cumulative hour; or

(2) When 50 percent of their useful life (or, for rechargeable batteries, 50 percent of their useful life of charge), as established by the transmitter manufacturer under [TSO-C91, paragraph (g)(2)], has expired.

The new expiration date for the replacement (or recharge) of the battery must be legibly marked on the outside of the transmitter and entered in the aircraft maintenance record. Subparagraph (d) (2) of this paragraph does not apply to batteries such as water-activated batteries) that are essentially unaffected during probable storage intervals.

(e) Notwithstanding paragraphs (a) and (b) of this section, a person may—

(1) Ferry a newly acquired airplane from the place where possession of it was taken to a place where the emergency locator transmitter is to be installed; and

(2) Ferry an airplane with an inoperative emergency locator transmitter from a place where repairs or replacement cannot be made to a place where they can be made.

No person other than required crewmembers may be carried board an airplane being ferried pursuant to paragraph (e) of this section.

§ 91.67 Right-of-way rules; except water operations.

(a) *General.* When weather conditions permit, regardless of whether an operation is conducted under Instrument Flight Rules or Visual Flight Rules, vigilance shall be maintained by each person operating an aircraft so as to see and void other aircraft in compliance with this section. When a rule of this section gives another aircraft the right of way, he shall give way to that aircraft and may not pass over, under, or ahead of it, unless well clear.

(b) *In distress.* An aircraft in distress has the right of way over all other air traffic.

(c) *Converging.* When aircraft of the same category are converging at approximately the same altitude (except head-on, or nearly so) the aircraft to the other's right has the right of way. If the aircraft are of different categories—

(1) A balloon has the right of way over any other category of aircraft;

(2) A glider has the right of way over an airship, airplane or rotorcraft; and

(3) An airship has the right of way over an airplane or rotorcraft.

However, an aircraft towing or refueling other aircraft has the right of way over all other engine-driven aircraft.

(d) *Approaching head-on.* When aircraft are approaching each other head-on, or nearly so, each pilot of each aircraft shall alter course to the right.

(e) *Overtaking.* Each aircraft that is being overtaken has the right of way and each pilot of an overtaking aircraft shall alter course to the right to pass well clear.

(f) *Landing.* Aircraft, while on final approach to land, or while landing, have the right of way over other aircraft in flight or operating on the surface. When two or more aircraft are approaching an airport for the purpose of landing, the aircraft at the lower altitude has the right of way, but it shall not take advantage of this rule to cut in front of another which is on final approach to land, or to overtake that aircraft.

(g) *Inapplicability.* This section does not apply to the operation of an aircraft on water.

§ 91.70 Aircraft speed.

(a) Unless otherwise authorized by the Administrator, no person may operate an aircraft below 10,000 feet MSL at an indicated airspeed of more than 250 knots (288 m.p.h.).

(b) Unless otherwise authorized or required by ATC, no person may operate an aircraft within an airport traffic area at an indicated airspeed of more than—

(1) In the case of a reciprocating engine aircraft, 156 knots (180 m.p.h.); or

(2) In the case of a turbine-powered aircraft, 200 knots (230 m.p.h.).

Paragraph (b) does not apply to any operations within a Terminal Control ARea. Such operations shall comply with paragraph (a) of this section.

(c) No person may operate an aircraft in the airspace underlying a terminal control area, or in a VFR corridor designated through a terminal control area, at an indicated airspeed of more than 200 knots (230 m.p.h.).

However, if the minimum safe airspeed for any particular operation is greater than the maximum speed prescribed in this section, the aircraft may be opened at that minimum speed.

§ 91.71 Acrobatic flight.

No person may operate an aircraft in acrobatic flight—

(a) Over any congested are of a city, town, or settlement:

(b) Over an open air assembly of persons:

(c) Within a control zone or Federal airway:

(d) Below an altitude of 1,500 feet above the surface: or

(e) When flight visibility is less than three miles.

For the purposes of this section, acrobatic flight means an intentional maneuver involving an abrupt change in an aircraft's attitude, an abnormal attitude, or abnormal acceleration, not necessary for normal flight.

§ 91.73 Aircraft lights.

No person may, during the period from sunset to sunrise (or, in Alaska, during the period a prominent unlighted object cannot be seen from a distance of three statute miles or the sun is more than six degrees below the horizon)—

(a) Operate an aircraft unless it has lighted position lights;

(b) Park or move an aircraft in, or in dangerous proximity to, a night flight operations area of an airport unless the aircraft—

(1) Is clearly illuminated;

(2) Has lighted position lights; or

(3) Is in an area which is marked by obstruction lights[.]

(c) Anchor an aircraft unless the aircraft—

(1) Has lighted anchor lights; or

(2) Is in an area where anchor lights are nor required on vessels[; or

[(d) Operate an aircraft, required by § 91.33 (c) (3) to be equipped with an anticollision light system, unless it has approved and lighted aviation red or aviation white anticollision lights. However, the anticollision lights need not be lighted when the pilot in command determines that, because of operating conditions, it would be in the interest of safety to turn the lights off.]

§ 91.75 Compliance with ATC clearances and instructions.

(a) When an ATC clearance has been obtained, no pilot in command may deviate from that clearance, except in an emergency, unless he obtains an amended clearance. However, except in positive controlled airspace, this paragraph does not prohibit him from cancelling an IFR flight plan if he is operating in VFR weather conditions. [If a pilot is uncertain of the meaning of an ATC clearance, he shall immediately request clarification from ATC.]

(b) Except in an emergency, no person may, in an area in which air traffic control is exercised, operate an aircraft contrary to an ATC instruction.

(c) Each pilot in command who deviates, in an emergency, from an ATC clearance or instruction shall notify ATC of that deviation as soon as possible.

(d) Each pilot in command who (though not deviating from a rule of this subpart) is given priority by ATC in an emergency, shall if requested by ATC, submit a detailed report of that emergency within 48 hours to the chief of that ATC facility.

§ 91.77 ATC light signals.

ATC light signals have the meaning shown in the following table [Fig. 5-36].

Color and type of signal	Meaning with respect to aircraft on the surface	Meaning with respect to aircraft in flight
Steady green	Cleared for takeoff	Cleared to land.
Flashing green	Cleared to taxi	Return for landing (to be followed by steady green at proper time).
Steady red	Stop	Give way to other aircraft and continue circling.
Flashing red	Taxi clear of runway in use.	Airport unsafe—do not land.
Flashing white	Return to starting point on airport.	Not applicable.
Alternating red and green.	Exercise extreme caution.	Exercise extreme caution.

Fig. 5-36. ATC light signals.

§ 91.79 Minimum safe altitudes; general.

Except when necessary for takeoff or landing, no person may operate an aircraft below the following altitudes:

(a) *Anywhere.* An altitude allowing, if a power unit fails, an emergency landing without undue hazard to persons or property on the surface.

(b) *Over congested areas.* Over any congested area of a city, town, or settlement, or over any open air assembly of persons, an altitude of 1,0000 feet above the highest obstacle within a horizontal radius of 2,000 feet of the aircraft.

(c) *Over other than congested areas.* An altitude of 500 feet above the surface, except over open water or sparsely populated areas. In that case, the aircraft may not be operated closer than 500 feet to any person, vessel, vehicle, or structure.

§ 91.81 Altimeter settings.

(a) Each person operating an aircraft shall maintain the cruising altitude or flight level of that aircraft, as the case may be, by reference to an altimeter that is set, when operating—

(1) Below 18,000 feet MSL, to—

(i) The current reported altimeter setting of a station along the route and within 100 nautical miles of the aircraft;

(ii) If there is no station within the area prescribed in subdivision (i) of this subparagraph, the current reported altimeter setting of an appropriate available station: or

(iii) In the case of an aircraft not equipped with a radio, the elevation of the departure airport or an appropriate altimeter setting available before departure; or

(2) At or above 18,000 feet MSL, to 29.92″ Hg.

§ 91.83 Flight plan; information required.

(a) Unless otherwise authorized by ATC, each person filing an IFR or VFR flight plan shall include in it the following information:

(1) The aircraft identification number and, if necessary, its radio call sign.

(2) The type of the aircraft or, in the case of a formation flight, the type of each aircraft and the number of aircraft, in the formation.

(3) The full name and address of the pilot in command or, in the case of a formation flight, the formation commander.

(4) The point and proposed time of departure.

(5) The proposed route, cruising

altitude (or flight level), and true airspeed at that altitude.

(6) The point of first intended landing and the estimated elapsed time until over that point.

(7) The radio frequencies to be used.

(8) The amount of fuel on board (in hours).

(9) In the case of an IFR flight plan, an alternate airport, except as provided in paragraph (b) of this section.

[(10) The number of persons in the aircraft, except where that information is otherwise readily available to the FAA.]

(11) Any other information the pilot in command or ATC believes is necessary for ATC purposes.

§ 91.85 Operating on or in the vicinity of an airport; general rules.

(a) Unless otherwise required by Part 93 of this chapter, each person operating an aircraft on or in the vicinity of an airport shall comply with the requirements of this section and of §§ 91.87 and 91.89.

(b) Unless otherwise authorized or required by ATC, no person may operate an aircraft within an airport traffic area except for the purpose of landing at, or taking off from, an airport within that area. ATC authorizations may be given as individual approval of specific operations or may be contained in written agreements between airport users and the tower concerned.

(c) After March 28, 1977, except when necessary for training or certification, the pilot in command of a civil turbojet-powered airplane shall use, as a final landing flap setting, the minimum certificated landing flap setting set forth in the approved performance information in the Airplane Flight Manual for the applicable conditions. However, each pilot in command has the final authority and responsibility for the safe operation of his airplane, and he may use a different flap setting approved for that airplane if he determines that it is necessary in the interest of safety.

§ 91.87 Operation at airports with operating control towers.

(a) *General.* Unless otherwise authorized or required by ATC, each person operating an aircraft to, from , or on an airport with an operating control tower shall comply with the applicable provisions of this section.

(b) *Communications with control towers operated by the United States.* No person may, within an airport traffic area, operate an aircraft to, from, or on an airport having a control tower operated by the United States unless two-way radio communications are maintained between that aircraft and the control tower. However, if the aircraft radio fails in flight, he may operate that aircraft and land if weather conditions are at or above basic VFR weather minimums, he maintains visual contact with the tower, and he receives a clearance to land. If the aircraft radio fails while in flight under IFR, he must comply with § 91.27.

(c) *Communications with other control towers.* No person may, within an airport traffic area, operate an aircraft to, from, or on an airport having a control tower that is operated by any person other than the United States unless—

(1) If that aircraft's radio equipment so allows, two-way radio communications are maintained between the aircraft and the tower; or

(2) If that aircraft's radio equipment allows only reception from the tower, the pilot has the tower's frequency monitored.

(d) *Minimum altitudes.* When operating to

an airport with an operating control tower, each pilot of—

(3) An airplane approaching to land on a runway served by a visual approach slope indicator, shall maintain an altitude at or above the glide slope until a lower altitude is necessary for a safe landing.

However, subparagraphs (2) and (3) of this paragraph do not prohibit normal bracketing maneuvers above or below the glide slope that are conducted for the purpose of remaining on the glide slope.

(e) *Approaches*. When approaching to land at an airport with an operating control tower, each pilot of—

(1) An airplane, shall circle the airport to the left; and

(2) A helicopter, shall avoid the flow of fixed-wing aircraft.

(f) *Departures*. No person may operate an aircraft taking off from an airport with an operating control tower except in compliance with the following:

(1) Each pilot shall comply with any departure procedures established for that airport by the FAA.

(2) Unless otherwise required by the departure procedure or the applicable distance from clouds criteria, each pilot of a turbine-powered airplane and each pilot of a large airplane shall climb to an altitude of 1,500 feet above the surface as rapidly as practicable.

(g) *Noise abatement runway system*. When landing or taking off from an airport with an operating control tower, and for which a formal runway use program has been established by the FAA, each pilot of a turbine-powered airplane and each pilot of a large airplane, assigned a noise abatement runway by ATC, shall use that runway. [However, consistent with the final authority of the pilot in command concerning the safe operation of the aircraft as prescribed in § 91.3(a), ATC may assign a different runway if requested by the pilot in the interest of safety.]

(h) Clearances required. No person may at an airport with an operating control tower, operate an aircraft on a runway or taxiway, or takeoff or land an aircraft, unless an appropriate clearance is received from ATC. A clearance to "taxi to" the takeoff runway assigned to the aircraft is not a clearance to cross that assigned takeoff runway, or to taxi on that runway at any point, but is a clearance to cross other runways that intersect the taxi route to that assigned takeoff runway. A clearance to "taxi to" any point other than an assigned takeoff runway is a clearance to cross all runways that intersect the taxi route to that point.]

§ 91.90 Terminal control areas.

(a) *Group I terminal control areas*.

(1) *Operating rules*. No person may operate an aircraft within a Group I terminal control area designated in Part 71 of this chapter except in compliance with the following rules:

(i) No person may operate an aircraft within a Group I terminal control area unless he has received an appropriate authorization from ATC prior to the operation of that aircraft in that area.

(ii) Unless otherwise authorized by ATC, each person operating a large turbine engine powered airplane to or from a primary airport shall operate at or above the designated floors while within the lateral limits of the terminal control area.

(2) *Pilot requirements*. The pilot in command of a civil aircraft may not land or take off that aircraft from an airport

within a Group I terminal control area unless he holds at least a private pilot certificate.

(3) *Equipment requirements.* Unless otherwise authorized by ATC in the case of in-flight VOR, TACAN, or two-way radio failure; or unless otherwise authorized by ATC in the case of a transponder failure occurring at any time, no person may operate an aircraft within a Group I terminal control area unless that aircraft is equipped with—

(i) An operable VOR or TACAN receiver (except in the case of helicopters);

(ii) An operable two-way radio capable of communicating with ATC on appropriate frequencies for that terminal control area; and

[(iii) The applicable equipment specified in § 91.24.]

(b) *Group II terminal control areas.*

(1) *Operating rules.* No person may operate an aircraft within a Group II terminal control area designated in Part 71 of this chapter except in compliance with the following rules:

(i) No person may operate an aircraft within a Group II Terminal Control Area unless he has received an appropriate authorization from ATC prior to operation of that aircraft in that area, and unless two-way radio communications are maintained, within that area, between that aircraft and the ATC facility.

(ii). Unless otherwise authorized by ATC, each person operating a large turbine engine powered airplane to or from a primary airport shall operate at or above the designated floors while within the lateral limits of the terminal control area.

(2) *Equipment requirements.* Unless otherwise authorized by ATC in the case of in-flight VOR, TACAN, or two-way radio failure; or unless otherwise authorized by ATC in the case of a transponder failure occurring at any time, no person may operate an aircraft within a Group II terminal control area unless that aircraft is equipped with—

(i) An operable VOR or TACAN receiver (except in the case of helicopters);

(ii) An operable two-way radio capable of communicating with ATC on the appropriate frequencies for that terminal control area; and

[(iii) The applicable equipment specified in § 91.24, except that automatic pressure reporting equipment is not required for any operation within the terminal control area, and a transponder is not required for IFR flights operating to or from an airport outside of but in close proximity to the terminal control area, when the commonly used transition, approach, or departure procedures to such airport require flight within the terminal control are.]

(c) *Group III terminal control areas.* [No person may operate an aircraft within a Group III Terminal Control Area designated in Part 71 unless the applicable provisions of § 91.24(b) are complied with, except that each compliance is not required if two-way radio communications are maintained, within the TCA, between the aircraft and the ATC facility, and the pilot provides position, altitude, and proposed flight path prior to entry.]

§ 91.97 Positive control areas and route segments.

(a) Except as provided in paragraph (b) of this section, no person may operate an aircraft within a positive control area or positive control route segment, designated in Part 71 of this chapter, unless that aircraft is—

(1) Operated under IFR at a specific flight level assigned by ATC;

(2) Equipped with instruments and equipment required for IFR operations:

(3) Flown by a pilot rated for instrument flight; and

(4) Equipped, when in a positive control area, with—

[(i) The applicable equipment specified in § 91.24; and]

(ii) A radio providing direct pilot controller communication on the frequency specified by ATC for the area concerned.

(b) ATC may authorize deviations from the requirements of paragraph (a) of this section. In the case of an inoperative transponder, ATC may immediately approve an operation within a positive control area allowing flight to continue, if desired, to the airport of ultimate destination, including any intermediate stops, or to proceed to a place where suitable repairs can be made, or both. A request for authorization to deviate from a requirement of paragraph (a) of this section, other than for operation with an inoperative transponder as outlined above must be submitted at least four days before the proposed operation in writing, to the ATC center having jurisdiction over the positive control area concerned. ATC may authorize a deviation on a continuing basis or for an individual flight, as appropriate.

VISUAL FLIGHT RULES

§ 91.105 Basic VFR Weather Minimums.

(a) Except as provided in § 91.107, no person may operate an aircraft under VFR when the flight visibility is less, or at a distance from clouds that is less, than that prescribed for the corresponding altitude in the following table: [Fig. 5-37].

Altitude	Flight visibility	Distance from clouds
1,200 feet or less above the surface (regardless of MSL altitude)—		
Within controlled airspace	3 statute miles	500 feet below. 1,000 feet above. 2,000 feet horizontal.
Outside controlled airspace	1 statute mile except as provided in § 91.105(b).	Clear of clouds.
More than 1,200 feet above the surface but less than 10,000 feet MSL—		
Within controlled airspace	3 statute miles	500 feet below. 1,000 feet above. 2,000 feet horizontal.
Outside controlled airspace	1 statue mile	500 feet below. 1,000 feet above. 2,000 feet horizontal.
More than 1,200 feet above the surface and at or above 10,000 feet MSL.	5 statute miles	1,000 feet below. 1,000 feet above. 1 mile horizontal.

Fig. 5-37. VFR weather minimums.

(b) When the visibility is less than one mile, a helicopter may be operated outside controlled airspace at 1,200 feet or less above the surface if operated at a speed that allows the pilot adequate opportunity to see any air traffic or other obstruction in time to avoid a collision.

(c) Except as provided in § 91.107, no person may operate an aircraft, under VFR, within a control zone beneath the ceiling when the ceiling is less than 1,000 feet.

(d) Except as provided in § 91.107, no person may take off or land an aircraft, or enter the traffic pattern of an airport, under VFR, within a control zone—

(1) Unless ground visibility at that airport is at least three statute miles; or

(2) If ground visibility is not reported at that airport, unless flight visibility during landing or take off, or while operating in the traffic pattern, is at least three statute miles.

(e) For the purposes of this section, an aircraft operating at the base altitude of a transition area or control area is considered to be within the airspace directly below that area.

§ 91.107 Special VFR weather minimums.

(a) Except as provided in § 93.113, when a person has received an appropriate ATC clearance, the special weather minimums of this section instead of those contained in § 91.105 apply to the operation of an aircraft by that person in a control zone under VFR.

(b) No person may operate an aircraft in a control zone under VFR except clear of clouds.

(c) No person may operate an aircraft (other than a helicopter) in a control zone under VFR unless flight visibility is at least one statute mile.

(d) No person may take off or land an aircraft (other than a helicopter) at any airport in a control zone under VFR—

(1) Unless ground visibility at that airport is at least one statute mile; or

(2) If ground visibility is not reported at that airport, unless flight visibility during landing or takeoff is at least one statute mile.

(e) No person may operate an aircraft (other than a helicopter) in a control zone under the special weather minimums of this section, between sunset and sunrise (or in Alaska, when the sun is more than six degrees below the horizon) unless:

(1) That person meets the applicable requirements for instrument flight under Part 61 of this chapter; and

(2) The aircraft is equipped as required in § 91.33(d).

[§ 91.09 VFR cruising altitude or flight level.

[Except while holding in a holding pattern of 2 minutes or less, or while turning, each person operating an aircraft under VFR in level cruising flight more than 3,000 feet above the surface shall maintain the appropriate altitude or flight level prescribed below, unless otherwise authorized b ATC:]

(a) When operating below 18,000 feet MSL and—

(1) On a magnetic course of zero degrees through 179 degrees, any odd thousand foot MSL altitude +500 feet (such as 3,500, 5,500, or 7,500); or

(2) On a magnetic course of 180 degrees through 359 degrees, any even thousand foot MSL altitude +500 feet (such as 4,500, 6,500, or 8,500).

(3) Degree to which the capability of the pilot to operate under IFR in the ATC system is impaired; and

(4) Nature and extent of assistance

he desires from ATC.

Subpart C—Maintenance, Preventive Maintenance, and Alterations

§ 91.161 Applicability.

(a) This subpart prescribes rules governing the maintenance, preventive maintenance, and alterations of U.S. registered civil aircraft operating within or without the United States.

(b) Sections 91.165, 91.169, 91.170, 91.171, 91.173, and 91.174 of this subpart do not apply to an aircraft maintained in accordance with a continuous airworthiness maintenance program as provided in Part 121, 1217, or 135 of this chapter.

§ 91.163 General.

(a) The owner or operator of an aircraft is primarily responsible for maintaining that aircraft in an airworthy condition, including compliance with part 39 of this chapter.

(b) No person may perform maintenance, preventive maintenance, or alterations on an aircraft other than as prescribed in this subpart and other applicable regulations, including Part 43.

[(c) No person may operate an aircraft for which a manufacturer's maintenance manual or Instructions for Continued Airworthiness has been issued that contains an Airworthiness Limitations section unless the mandatory replacement times, inspection intervals, and related procedures specified in that section or alternative inspection intervals and related procedures set forth in an operations specification approved by the Administrator under Parts 121, 123, 127, or 135, or in accordance with an inspection program approved under § 91.217(e), have been complied with.]

§ 91.165 Maintenance required.

Each owner or operator of an aircraft shall have that aircraft inspected as prescribed in Subpart D or § 91.169 of this Part, as appropriate, and § 91.170 of this Part and shall, between required inspections, have defects repaired as prescribed in Part 43 of this chapter. [In addition, each owner or operator shall ensure that maintenance personnel make appropriate entries in the maintenance records indicating that the aircraft has been approved for return to service.]

§ 91.167 Carrying persons other than crewmembers after repairs or alterations.

(a) No person may carry any person (other than crewmembers) in an aircraft that has been repaired or altered in a manner that may have appreciably changed its flight characteristics, or substantially affected its operation in flight, until it has been approved for return to service in accordance with Part 43 and an appropriately rated pilot, with at least a private pilot's certificate, flies the aircraft, makes an operational check of the repaired or altered part, and logs the flight in the aircraft's records.

(b) Paragraph (a) of this section does not require that the aircraft be flown if ground tests or inspections, or both, show conclusively that the repair of alteration has not appreciably changed the flight characteristics, or substantially affected the flight operation of the aircraft.

§ 91.169 Inspections.

(a) Except as provided in paragraph (c) of this section, no person may operate an aircraft unless, within the preceding 12 calendar months, it has had—

(1) An annual inspection in accordance with Part 43 of this chapter and has been approved for return to service by a person authorized by § 43.7 of this chapter; or

(2) An inspection for the issue of an airworthiness certificate.

No inspection performed under paragraph

(b) of the section may be substituted for any inspection required by this paragraph unless it is performed by a person authorized to perform annual inspections, and is entered as an "annual" inspection in the required maintenance records.

(b) Except as provided in paragraph (c) of this section, no person may operate an aircraft carrying any person (other than a crewmember) for hire, and no person may give flight instruction for hire in an aircraft which that person provides, unless within the preceding 100 hours of time in service it has received an annual or 100-hour inspection and been approved for return to service in accordance with Part 43 of this chapter, or received an inspection for the issuance of an airworthiness certificate in accordance with Part 21 of this chapter. The 100-hour limitation may be exceeded by not more than 10 hours if necessary to reach a place at which the inspection can be done. The excess time, however, is included in computing the next 100 hours of time in service.

(c) Paragraphs (a) and (b) of this section do not apply to—

(1) Any aircraft for which its registered owner or operator complies with the progressive inspection requirements of § 91.171 and Part 43 of this chapter;

(2) An aircraft that carries a special flight permit or a current experimental or provisional certificate;

(3) Any airplane operated by an air travel club that is inspected in accordance with Part 123 of this chapter and the operator's manual and operations specifications; or

(4) An aircraft inspected in accordance with an approved aircraft inspection program under Part 135 of this chapter and so identified by the registration number in the operations specifications of the certificate holder having the approved inspection program.

(5) Any large airplane, or a turbojet or turbopropeller-powered multiengine airplane, that is inspected in accordance with an inspection program authorized under Subpart D of this Part.

§ 91.173 Maintenance records.

(a) Except for work performed in accordance with [§ 91.171], each registered owner or operator shall keep the following records for the periods specified in paragraph (b) of this section:

(1) Records of the maintenance and alternation, and records of the 100-hour, annual, progressive, and other required or approved inspections, as appropriate, for each aircraft (including the airframe) and each engine, propeller, rotor, and appliance of an aircraft. The records must include—

(i) A description (or reference to data acceptable to the Administrator) of the work performed;

(ii) The date of the completion of the work performed; and

(iii) The signature and certificate number of the person approving the aircraft for return to service.

(2) Records containing the following information:

(i) The total time in service of the airframe, each engine and each propeller.

(ii) The current status of life-limited parts of each airframe, engine, propeller, rotor, and appliance.

(iii) The time since last overhaul of items installed on the aircraft which are required to be overhauled on a specified time basis.

(iv) The identification of the current inspection status of the aircraft, including the times since the last inspections required by the inspection program under which the aircraft and its appliances are maintained.

(v) The current status of applicable airworthiness directives (AD) including, for each, the method of compliance, the AD number, and revision date. If the AD involves recurring action, the time and date when the next action is required.

[(vi) Copies of the forms prescribed by § 43.9(a) of this chapter for each major alteration to the airframe and currently installed engines, rotors, propellers, and appliances.]

(b) The owner and operator shall retain the following records for the periods prescribed:

(1) The records specified in paragraph (a) (1) of this section shall be retained until the work is repeated or superseded by other work or for one year after the work is performed.

(2) The records specified in paragraph (a) (2) of this section shall be retained and transferred with the aircraft at the time the aircraft is sold.

(3) A list of defects furnished to a registered owner or operator under [§ 43.11] of this chapter, shall be retained until the defects are repaired and the aircraft is approved for return to service.

(c) The owner or operator shall make all maintenance records required to be kept by this section available for inspection by the Administrator or any authorized representative of the National Transportation Safety Board (NTSB).

[(Approved by the Office of Management and Budget under OMB control number 2120-0005)]

Title 49–Transportation

National Transportation Safety Board

REVISED: SEPTEMBER 11, 1980

PART 830—NOTIFICATION AND REPORTING OF AIRCRAFT ACCIDENTS OR INCIDENTS AND OVERDUE AIRCRAFT, AND PRESERVATION OF AIRCRAFT WRECKAGE, MAIL, CARGO, AND RECORDS.

Subpart A—General

§ 830.1 Applicability.

This part contains rules pertaining to:

(a) Notification and reporting aircraft accidents and incidents and certain other occurrences in the operation of aircraft when they involve civil aircraft of the United States wherever they occur, or foreign civil aircraft when such events occur in the United States, its territories or possessions.

(b) Preservation of aircraft wreckage, mail, cargo, and records involving all civil aircraft in the United States, its territories or possessions.

§ 830.2 Definitions.

As used in this part the following words or phrases are defined as follows:

"Aircraft accident" means an occurrence associated with the operation of an aircraft which takes place between the time any person boards the aircraft with the intention of flight and all such persons have disembarked, and in which any person suffers death or serious injury, or in which the aircraft receives substantial damage.

"Fatal injury" means any injury which results in death within 30 days of the accident.

"Incident" means an occurrence other than an accident, associated with the operation of an aircraft, which affects or could affect the safety of operations.

"Operator" means any person who causes or authorizes the operation of an aircraft, such as the owner, lessee, or bailee of an aircraft.

"Serious injury" means any injury which (1) requires hospitalization for more than 48 hours, commencing within 7 days from the date of the injury was received; (2) results in a fracture of any bone (except simple fractures of fingers, toes, or nose); (3) causes severe hemorrhages, nerve, muscle, or tendon damage; (4) involves any internal organ; or (5) involves second- or third-degree burns, or any burns affecting more than 5 percent of the body surface.

"Substantial damage" means damage or failure which adversely affects the structural strength, performance, or flight characteristics of the aircraft, and which would normally require major repair or replacement of the affected component. Engine failure, damage limited to an engine, bent fairings or cowling, dented skin, small punctured holes in the skin or fabric, ground damage to rotor or propeller blades, damage to landing gear, wheels, tires, flaps, engine accessories, brakes, or wingtips are not considered "substantial damage" for the purpose of this part.

Subpart B—Initial Notification of Aircraft Accidents, Incidents, and Overdue Aircraft

§ 830.5 Immediate notification.

The operator of an aircraft shall immediately, and by the most expeditious means available, notify the nearest National Transportation Safety Board (Board), field office[1] when:

(a) An aircraft accident or any of the following listed incidents occur:

(1) Flight control system malfunction or failure;

(2) Inability of any required flight crewmember to perform normal flight duties as a result of injury or illness;

(3) Failure of structural components of a turbine engine excluding compressor and turbine blades and vanes;

(4) In-flight fire; or

(5) Aircraft collide in flight.

(b) An aircraft is overdue and is believed to have been involved in an accident.

§ 830.6 Information to be given in notification.

The notification required in § 830.5 shall contain the following information, if available:

(a) Type, nationality, and registration marks of the aircraft;

(b) Name of owner, and operator of the aircraft;

(c) Name of the pilot-in-command;

(d) Date and time of the accident;

(e) Last point of departure and point of intended landing of the aircraft;

(f) Position of the aircraft with reference to some easily defined geographical point;

(g) Number of persons aboard, number killed, and number seriously injured;

(h) Nature of the accident, the weather and the extent of damage to the aircraft, so far as is known; and

(i) A description of any explosives, radioactive materials, or other dangerous articles carried.

[1]The National Transportation Safety Board field offices are listed under U.S. Government in the telephone directories in the following cities: Anchorage, Alaska; Atlanta, Ga.; Chicago, Ill.; Denver, Colo.; Fort Worth, Tex.; Kansas City, Mo.; Los Angeles, Calif.; Miami, Fla.; New York, N.Y.; Seattle, Wash.

Subpart C—Preservation of Aircraft Wreckage, Mail, Cargo, and Record

§ 830.10 Preservation of aircraft wreckage, mail, cargo, and records.

(a) The operator of an aircraft involved in an accident or incident for which notification must be given is responsible for preserving to the extent possible any aircraft wreckage, cargo, and mail aboard the aircraft, and all records, including all recording mediums of flight, maintenance, and voice recorders, pertaining to the operation and maintenance of the aircraft and to the airmen until the Board takes custody thereof or a release is granted pursuant to § 831.10(b).

(b) Prior to the time the Board or its authorized representative takes custody of aircraft wreckage, mail, or cargo, such wreckage, mail, or cargo may not be disturbed or moved except to the extent necessary:

(1) To remove persons injured or trapped;

(2) To protect the wreckage from further damage; or

(3) To protect the public from injury.

(c) Where it is necessary to move aircraft wreckage, mail or cargo, sketches, descriptive notes, and photographs shall be made, if possible, of the original position and condition of the wreckage and any significant impact marks.

(d) The operator of an aircraft involved in an accident or incident shall retain all records, reports, internal documents, and memorands dealing with the accident or incident, until authorized by the Board to the contrary.

Subpart D—Reporting of Aircraft Accidents, Incidents, and Overdue Aircraft

§ 830.15 Reports and statements to be filled.

(a) *Reports.* The operator of an aircraft shall file a report on Board Form 6120.1 or Board Form 6120.2 within 10 days after an accident, or after 7 days if an overdue aircraft is still missing. A report on an incident for which notification is required by § 830.5(a) shall be filed only as requested by an authorized port on an incident for which notification is required by § 830.5(a) shall be filed only as requested by an authorized representative of the Board.

(b) *Crewmember statement.* Each crewmember, if physically able at the time the report is submitted, shall attach a statement setting forth the facts, conditions, and circumstances relating to the accident or incident as they appear to him. If the crewmember is incapacitated, he shall submit the statement as soon as he is physically able.

(c) *Where to file the reports.* The operator of an aircraft shall file any report with the field office of the Board nearest the accident or incident.

Note.—The reporting and recordkeeping requirements contained herein have been approved by the Office of Management and Budget in accordance with the Federal Report Act of 1942.

Chapter 6

Private Pilot Flight Maneuvers

This chapter is devoted to all of the nifty maneuvers and exercises that you'll be doing with your flight instructor. The maneuvers are discussed here to help you understand the procedures and theories of each. You should also check into your *Private Pilot Flight Test Guide* (AC 61-54A for Airplanes) to familiarize yourself with the FAA standards for each maneuver. Remember that basic pattern work for takeoffs and landings was covered in Chapter 2 for the pre-solo level.

HOW THE AIRPLANE TURNS

In Fig. 6-1A, the airplane is maintaining its altitude because the total lift force is 1600 pounds and is acting vertically, opposing the total weight of 1600 pounds. There is no horizontal component of lift, so the aircraft will not turn. Now put the aircraft into a 30 degree bank without applying back pressure to the elevator control, as shown in B of Fig. 6-1. The total lift is divided into a horizontal component of 800 pounds and a vertical component of 1380 pounds. The vertical component of lift is now less than the weight of the aircraft; this means it will lose altitude. Obviously, to maintain the same altitude, we must increase the vertical component of lift to equal the weight of the aircraft. The best means available to accomplish this is to increase the total lift force by applying up elevator to raise the angle of attack.

When the total lift is increased until the vertical component equals the weight, the aircraft will maintain altitude as in Fig. 6-1C. The specific numerical values of this example are not important, but it *is* important that the student understands why the back pressure is needed in a level, banked turn. Obviously we don't have any way to determine if the vertical

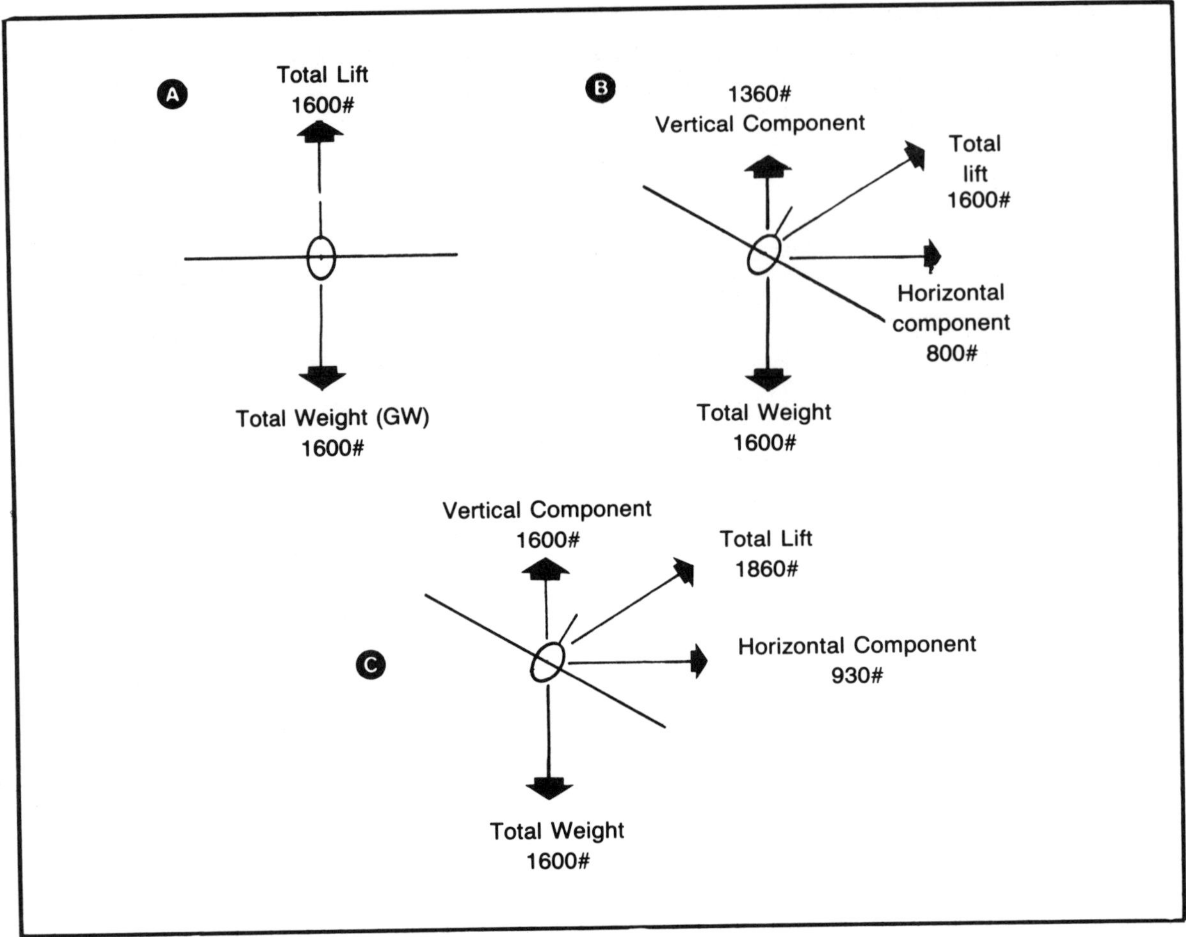

Fig. 6-1. How the airplane turns.

component of lift equals the weight other than the VSI or altimeter. The student will soon learn to feel how much back pressure is required for the bank used and utilize his instruments only as a check. As the amount of bank increases, the back pressure will also increase to maintain the altitude, and as the bank decreases, the back pressure will also decrease to nothing in a wings-level attitude. It can be seen from the example in Fig. 6-2 that the horizontal component of lift is responsible for turning the airplane.

To take it one step further, we can show why there is a 2G force in a 60 degree banked level turn. In Fig. 6-2(A), the lift force of 1600 pounds is acting vertically against the weight of the aircraft. In Fig. 6-2B, the aircraft is in a 60 degree bank with no back pressure applied and now the vertical component of lift is only 800 pounds, or half the weight of the aircraft. The pilot's job is to increase the total amount of lift on the wings to the point where the vertical component of lift will equal the weight of the aircraft.

To do this, we must double the total lift (Figure 6-2C), which will result in the 2G force on the entire aircraft and its occupants. Beyond a 60 degree bank, the load factors or G forces increase rapidly. You should avoid excessively steep banks because of the high stress it places on the aircraft. Since there is more horizontal lift in this steeper bank, the rate of turn will be greater and the radius will be less than in a 30 degree bank.

GROUND REFERENCE MANEUVERS

The primary reason for ground reference maneuvers is to show you the effects of wind on the ground track of the airplane and how to correct for the wind to obtain a desired ground track. They also teach you to fly the aircraft while your attention is divided between the ground and the airplane. Proficiency in ground reference maneuvers is necessary before a proper traffic pattern can be flown. We determine the wind at altitude by a *wind circle*—a constant-banked turn around an intersection to show clearly the downwind drift of the airplane.

The rectangular pattern shown in Fig. 6-3 has a wind from the south at 20 mph and the aircraft has a true airspeed of 100 mph. Our

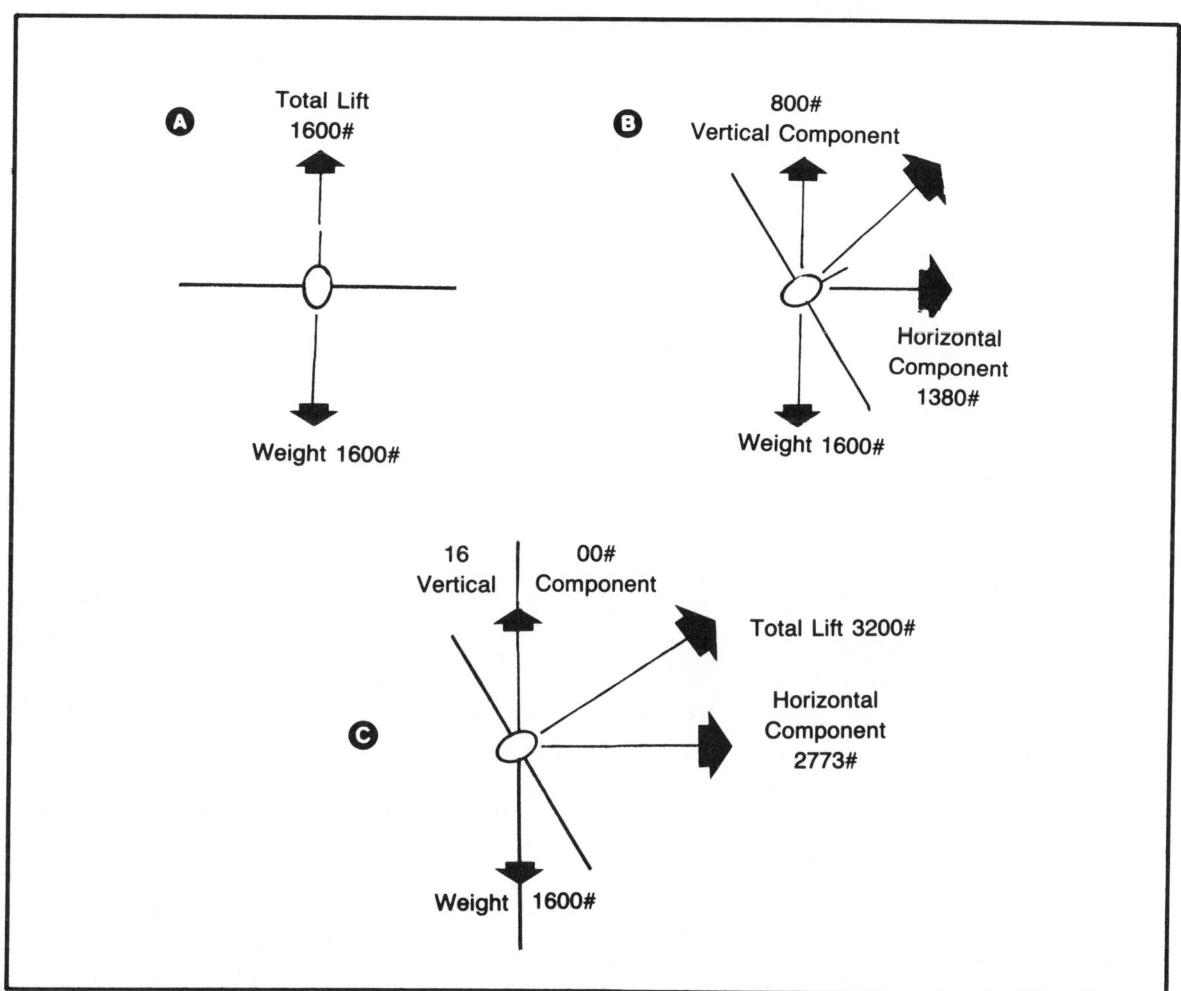

Fig. 6-2. Turning forces and components.

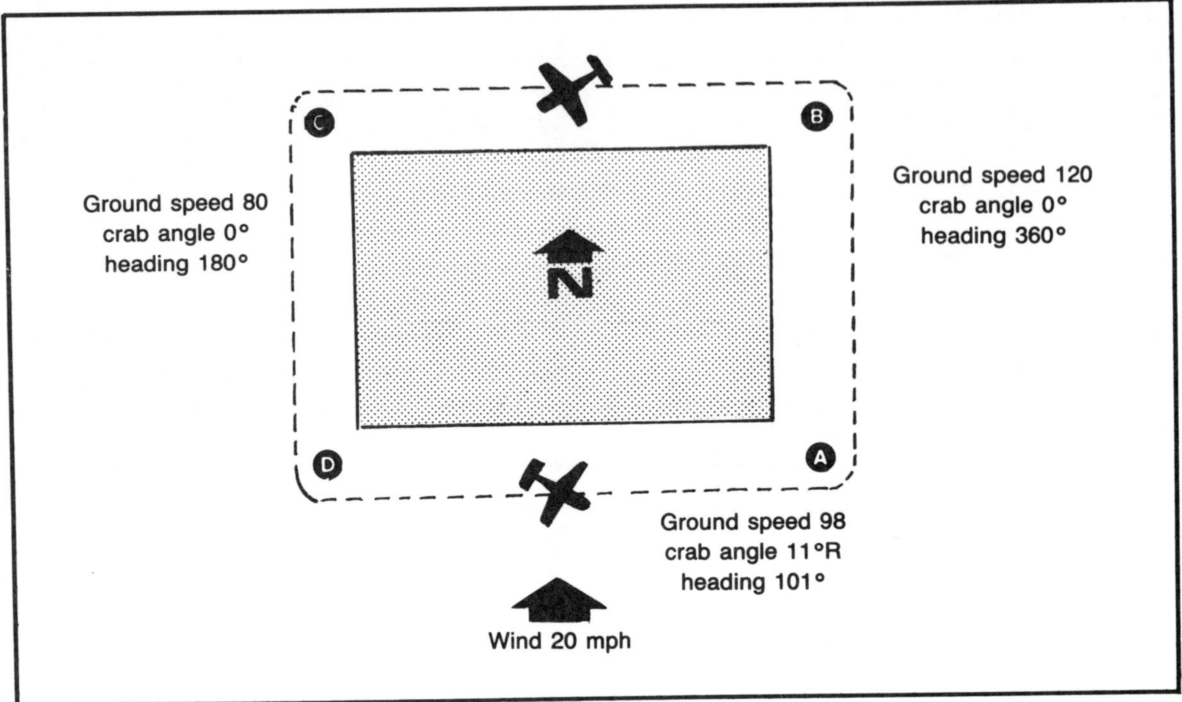

Fig. 6-3. Rectangular course.

objective is to fly around the field, keeping a constant distance from each of the four sides, and at a constant altitude. This maneuver should be flown at pattern altitude, usually 800 feet AGL. This altitude is low enough that you can see the effects of the wind, but not so low as to violate the regulations about staying a safe altitude above persons or property on the ground (500′). The distance from the field should be such that it appears at the same angle to the student as the runway would on downwind. It is most important that the student understand where the numerical values for ground speed and heading change and why they change, as they would apply to flying the aircraft.

It is fairly obvious that when flying from point A to B, the ground speed will be the fastest and there will be no *crab* (wind correction angle), since there is a direct tailwind. This also means that during the turn at point B, Fig. 6-3, the least amount of time will be available and the turn will take more than 90 degrees of heading change to compensate for the wind from point B to point C. Since we have less time and more heading change, the bank must be the steepest (bank governs rate of turn if airspeed is constant). From B to C, the ground speed is less because the wind is now from the side and a crab is needed into the wind to maintain the desired ground track.

The turn at point C will need a lesser amount of bank for two reasons: The ground speed is slower, which allows more time, and the amount of heading change is less than 90 degrees. The amount of bank used in the turns is acquired by experience and judgement. As the bank varies, the back pressure must also vary to maintain a constant altitude. The rectangular pattern will be done first with left turns, but right turns are also included after you get the idea of how to correct for the wind.

The techniques of a rectangular pattern apply directly to the landing pattern. After all, the runway is on the ground and when we turn final, we want to be lined up directly with the center line. We can vary our heading and bank in the landing pattern to accomplish this with the only difference being climbing and descending while making the turns. It is important to note that as the airspeed decreases, the effects of the wind will become more pronounced, so as you slow the aircraft to approach speed on the downwind leg, more crab will be needed to maintain a straight path over the ground.

S-Turns

S-turns across a road are good maneuver for getting used to dividing your attention between the airplane and the ground, although their primary purpose is to show you how to correct for wind in a turn. There was an introduction to this idea in the rectangular course, but S-turns give you a chance to acquire finesse. S-turns are a series of 180 degree turns of about a quarter-mile radius using a road as nearly perpendicular to the wind as possible. The object is to fly a series of semicircles of the same size, making smooth, coordinated turns to correct for wind drift by varying the steepness of the bank. The aircraft should cross the road in a level attitude with the longitudinal axis perpendicular to the road.

In Fig. 6-4, the maneuver is entered downwind because this will require the steepest initial bank due to the high ground speed. As we progress around the turn and begin to crab, the ground speed slows and therefore the bank must get shallower. If the bank did not shallow, we would be turning at the same number of degrees per second, but as we began to head into the wind, the ground speed would drop and the aircraft would not follow the smooth curve of the semicircle. The shallowing of the turn should be such that the wings are level as the aircraft crosses the road at point C.

After crossing the road, the bank should be a shallow one in the opposite direction. When point D is reached, the bank must be steepened in order to have the turn completed as the aircraft crosses the road again. The series can be continued until you get tired or run out of road. Remember, the greater the ground speed, the greater the angle of bank.

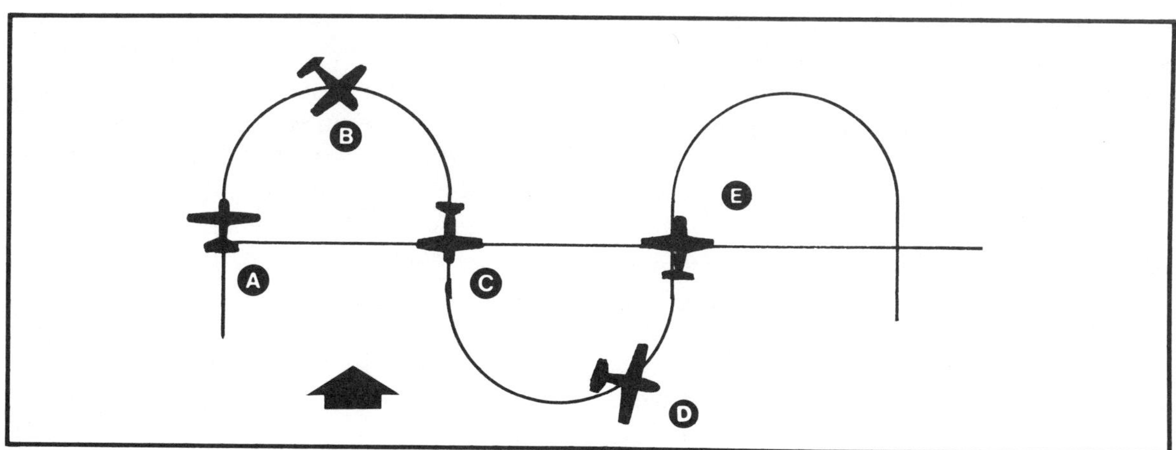

Fig. 6-4. S-turns.

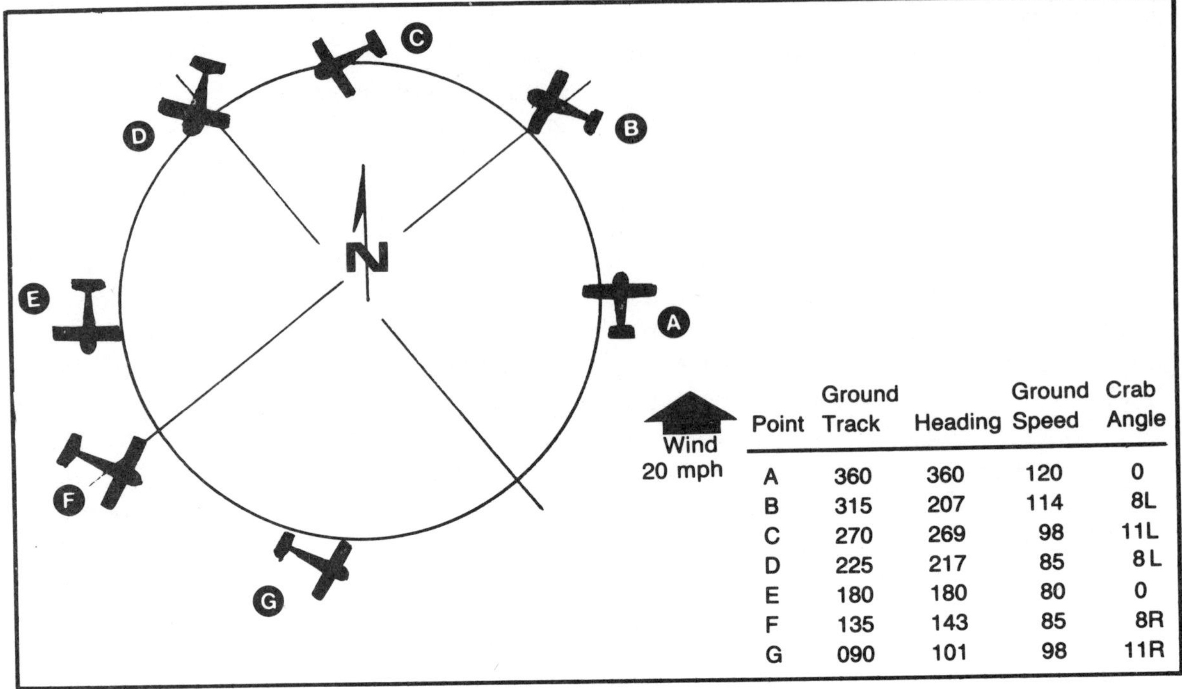

Point	Ground Track	Heading	Ground Speed	Crab Angle
A	360	360	120	0
B	315	207	114	8L
C	270	269	98	11L
D	225	217	85	8L
E	180	180	80	0
F	135	143	85	8R
G	090	101	98	11R

Fig. 6-5. Turns about a point.

Turns about a Point

Again, as in all the examples with numbers, the specific values are not important. The important thing is to understand where the values *change* and where the *greatest* changes occur. Since altitude and airspeed are going to remain the same for all practical purposes, the only correction available to us will be the amount of bank. Bank governs the rate of turn or how fast the aircraft is changing heading.

Starting at point A and going to point B, in Fig. 6-5, notice the change in heading must be the greatest (106 degrees) and we have the least amount of time to do it in since our ground speed is greatest at point A. This then means that our steepest bank will be at point A. Between point B and D, the bank will be shallower because we have more time to turn the aircraft since our ground speed is slower and requires a smaller amount of heading change, that is, 90 degrees. From point D to F, the bank angle will be the least because this is the slowest ground speed (80) and requires the least amount of heading change (74 degrees). This means that we have the longest amount of time to change heading the least, which results in the shallowest bank (in other words, the slower rate of turn.) From point F to H, the bank will steepen up again because the amount of heading change required is increasing (90 degrees) and the time to change the heading that much is getting shorter due to the increasing ground speed (90 mph).

The bank required to fly turns about a point is determined by two factors: The ground speed, which is going to determine how much time is available, and the crab angle, which determines how much the heading must change from point to point. Both factors are equally important.

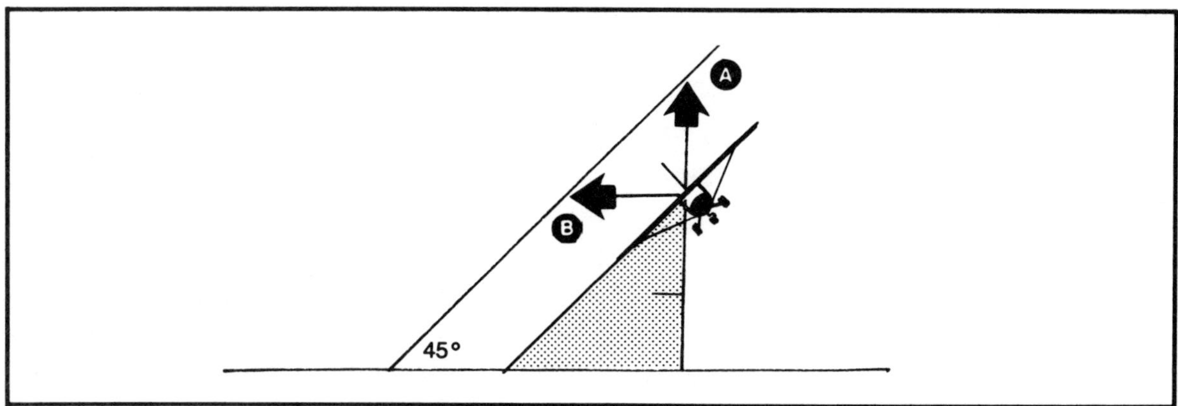

Fig. 6-6. Vision obstruction in a 45-degree bank.

This maneuver is normally entered with the aircraft heading directly downwind because this is where the steepest bank is required. If we enter downwind and roll into a bank of 45 degrees opposite our point, we can immediately determine if our distance and altitude are correct for the rest of the maneuver.

The aircraft is in a 45 degree bank but the point is hidden from the pilot (Fig. 6-6). He/she can either climb the aircraft to point A, move closer to the point as at point B, or do a combination of both.

The four parts of Fig. 6-7 show what the point will look like to the pilot as the aircraft moves around the point. Notice that the point moves (in relation to a fixed point on the wing) as the bank and crab angles change. Starting at point A, there is no crab and the bank is the steepest, so the point is directly under the cen-

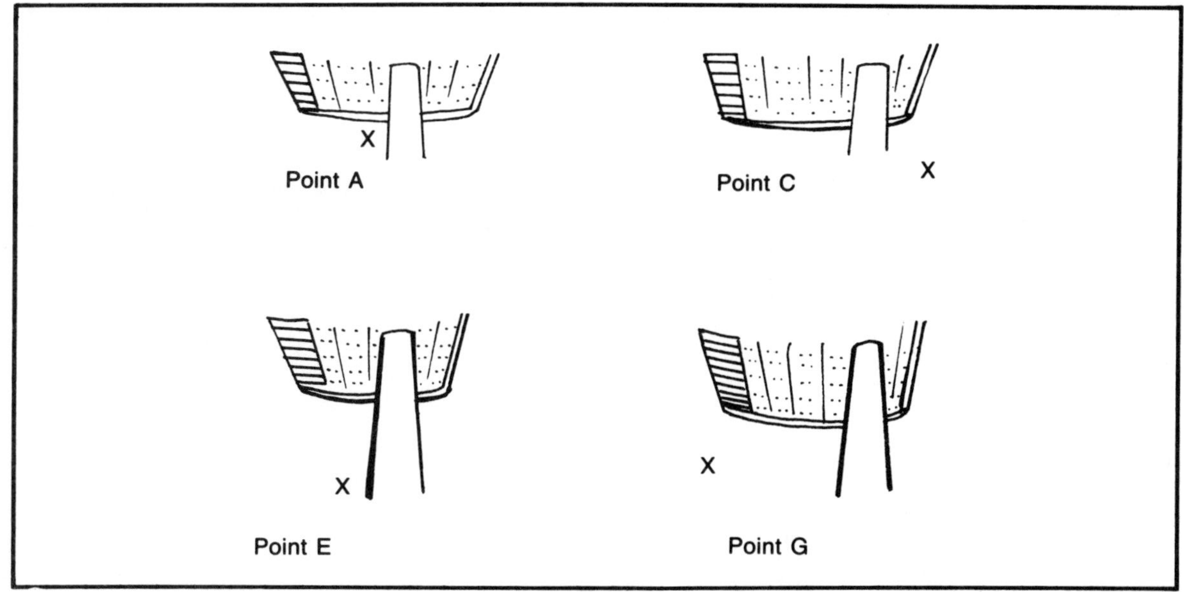

Fig. 6-7. Turns about a point—cockpit visual reference.

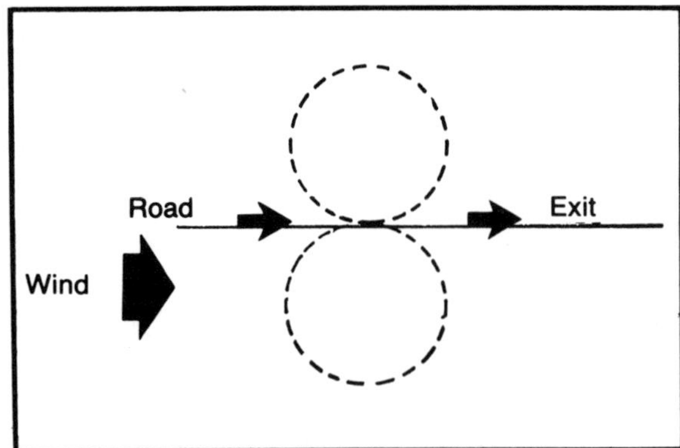

Fig. 6-8. Eights along a road.

ter of the wing. Moving around the turn to point C, the bank is lessened, which raises the wing about the point and maximum crab is present, which will move the point in front of the wing. At point E, there is no crab so the point is directly under the center of the wing again; now the bank is at its least so the wing will be significantly higher above the point. At point G, the bank is increasing again so the wing is moving downward toward the point and the crab is greatest (in the other direction), so the point will be behind the wing.

Usually we will continue around the point for a second turn and roll out when we reach point A the second time. The numerical values of crab and bank are determined entirely by the airspeed of the airplane and the velocity of the wind. In actual practice, the bank variance (other than the initial 45 degree bank) is determined by the pilot's judgement to remain a constant distance from the point. If there is no wind, the bank would stay at 45 degrees for the two turns.

Time Is Money!

Something should be mentioned about minimizing wasted time in setting up for any ground reference maneuver. First, descend to your desired altitude using outside reference rather than the altimeter. After reaching your altitude you can use the altimeter to maintain that height above the ground while you determine the wind direction with a wind circle. Then turn the airplane downwind, pick the point, and enter. The usual tendency is to do a lot of maneuvering to get just right on a specific point, but a lot of unproductive time can be eliminated by using any prominent feature. So to review briefly, get your altitude, find the wind, turn downwind, pick the point, and enter.

The Eights Maneuvers

There are three basic "eights" maneuvers; eights along a road, eights across a road, and eights around pylons.

First let's take the eights *along* a road (Fig. 6-8). These are entered with the aircraft tracking along the road and the sides of the eight are on either side of the road, both having the same radius.

Second, we have the eights *across* a road. These are entered with the aircraft tracking *across* the road at a 45 degree angle to it and flying downwind (Fig. 6-9).

And finally, the eights around pylons are nothing more than two symmetrical circles around two different points on the ground. The entry is downwind as shown in Fig. 6-10.

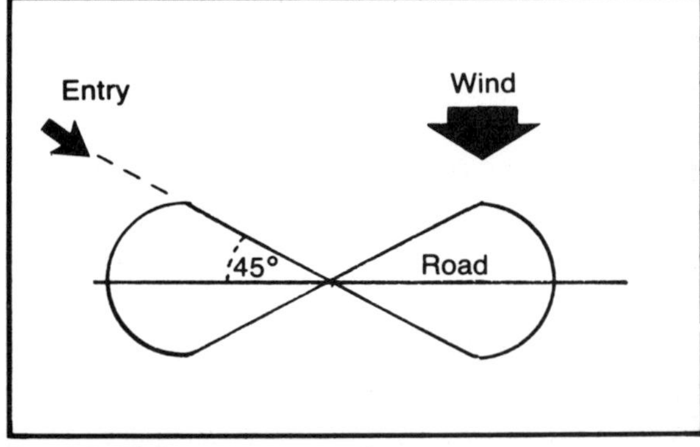

Fig. 6-9. Eights across a road.

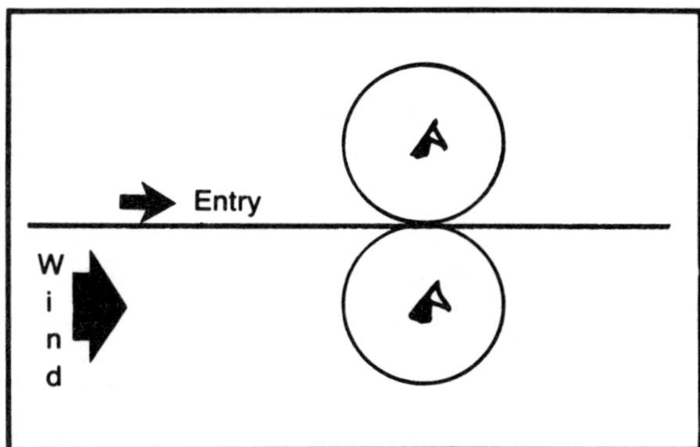

Fig. 6-10. Eights around pylons.

STALLS

First of all, a stall has nothing to do with the engine as one would think in trying to relate to an automobile. A stall is caused by an excessive angle of attack, and nothing more. Therefore, to recover from a stall, all that is needed is to reduce the angle of attack by releasing back pressure and the airplane will fly again. When the wing reaches a stalled condition, it is no longer producing lift, so we will see it as the nose dropping. With a nose down-attitude, and the back pressure released, the aircraft will accelerate out of the stall condition by itself. Stalls are practiced for two reasons: to recognize the approach to a stall, and to recover from the stall with minimum loss of altitude. There are three basic stalls that will be practiced; the *approach to landing stall,* the *takeoff and departure stall,* and the *accelerated stall.*

Approach to Landing Stall

The approach to landing stall is begun as if the aircraft is on the downwind leg of the traffic pattern. Carb heat is applied, but the altitude is maintained and power will be reduced to approach power (i.e., 1700 rpm), and the aircraft is slowed to approach speed and configuration (flaps) as in a normal approach. Two clearing turns are done, which simulate base and final. After the aircraft is trimmed for the final approach speed and configuration, the nose is brought up until the stall occurs. When the student recognizes the stall, back pressure is released, full throttle is applied, and carb heat returned to the OFF position. Right rudder must also be applied to counteract the torque and P-factor as power is applied. If flaps were used, they should be immediately retracted to 20 degrees to minimize drag. This is where timing the flaps down and up during the run-up checks will pay off.

Now care is taken to minimize the loss of altitude while accelerating the aircraft to the best-angle-of-climb speed (V_x) before trying to climb. To make this seemingly complicated procedure easier to understand, think of it as setting the aircraft up just exactly as if you were going to land. In fact, this is why it is called the approach to landing stall. They should be done straight ahead and with 20 degree banks both left and right, with and without flaps. Since a normal landing is considered to be with full flaps, toward the end of the private course, all your approach to landing stalls should be done with full flaps, even though you may still be limited to 20 degrees of flaps for your landings.

Takeoff and Departure Stalls

Takeoff and departure stalls, as the name implies, will simulate an accident after liftoff and close to the ground. The easiest way to avoid departure stalls is to remain in ground effect until the best rate-of-climb speed (V_y) is reached before starting a climb. Unfortunately, this type of stall is one of the most prevalent accident causes today.

To practice this maneuver, we will again do our two clearing turns and have the aircraft in the takeoff configuration and speed at the completion of the second turn. During the first clearing turn, carb heat is applied while still maintaining altitude; during the second turn, sufficient power is reduced to slow the aircraft to rotation speed while still maintaining altitude. As we roll out of the second clearing turn at rotation speed, full power is again applied and carb heat taken off.

At this time the aircraft is in the same attitude and configuration as when it lifts off the runway. Now, instead of lowering the nose and allowing the aircraft to accelerate, we will increase the pitch even more to an attitude that will cause the aircraft to stall. More and more right rudder is applied as the aircraft slows down, to overcome torque and P-factor. Again, coordinated recovery techniques are used. Back pressure is released to break the stall and coordinated aileron and rudder are used to level the wings. Minimum loss of altitude is again important while allowing the aircraft to accelerate to its best angle of climb speed. These will be practiced straight ahead and with 20 degree banks both right and left.

Accelerated Stalls

There are two types of accelerated stalls that should be demonstrated, but only one need be practiced for the flight test. They both are used to show that the aircraft can be stalled in any altitude, even level flight.

The "demonstrated only" version is entered as a power-off stall, clean configuration, straight ahead. The nose is brought up to what you normally think of as a stall attitude, but before it stalls, lower the nose to substantially nose-down attitude to gain a little airspeed. Then the elevator is smoothly but rapidly pulled all the way back. As the aircraft returns to level pitch, it will momentarily stall with considerable buffeting. Recovery is the same; release back pressure and add power to stop the descent and accelerate back to cruise airspeed.

With this maneuver it is easy to see the change of angle of attack. The aircraft is started down with a nose-down attitude. Then the attitude is changed without changing the direction of flight, which results in a high angle of attack and thus the stall in a level flight attitude.

The accelerated stall that you will do in training is done with reduced power and a 45 degree bank. After the two clearing turns and power reduction to about 2200 rpm, the pilot attempts to hold altitude with elevator only. Since there is insufficient power available, the aircraft will stall instead. If the aircraft is coordinated when it stalls, the high wing will stall first and when back pressure is released, the aircraft is flying again. As usual, we will try to recover with minimal loss of altitude.

In a turn, the aircraft usually tends to stall the top wing first. If the maneuver is perfectly coordinated, and the aircraft is in perfect rig, the top wing stalls first because just prior to the stall, the aircraft starts to slide downward slightly, toward the low wing. The fuselage then blocks the airflow over the top wing and it will stall.

Common Pilot Errors In Stalls

The most common pilot errors during the stalls are:

1. Not compensating sufficiently for the overbanking tendency in steep turns so excessive bank is developed in the accelerated stall maneuver, which normally results in the second error:

2. Allowing the pitch attitude to decrease so the aircraft continues to fly and the maneuver turns into a steep spiral. If this happens, it is probably best to just start over.

3. Subconsciously trying to hold the nose up with top rudder! This results in a slip and the aircraft doesn't want to stall while in a slip.

Invariably with power-on stalls the student gets apprehensive about the maneuver, feeling that it can't be controlled. Nothing could be further from the truth! All stalls are controlled directly by the pilot *whether or not* he is aware of it.

Let's talk about straight-ahead power-on stalls. As the pitch attitude is raised to the desired height, the airspeed decays and torque and P-factor become more pronounced. This makes the aircraft roll and yaw to the left unless we compensate with enough right rudder.

The big question is: How much is enough? Pay particular attention to the bank and pitch attitude out the windshield. Most people tend to stare at the ball and never notice what the aircraft is doing. It is especially important to freeze the ailerons in the neutral position and then maintain heading and bank with the rudder only. If we keep loose on the rudders and keep working them to do the desired job, when the aircraft stalls, it will be straight ahead. However, if we have too much right rudder or too much left rudder, the fuselage will blank out part of the airflow over that wing and it will stall first. So if we want the right wing to stall first, hold too much right rudder, and if we want the left wing to stall first, hold too much left rudder. Notice that in either case, the ball will be deflected away from the direction of yaw and the aircraft will roll away from the ball.

Another common tendency is for the student to hold some right aileron to keep the aircraft from turning. Right aileron will result in the left wing stalling first for two reasons: Insufficient right rudder is used so the left wing gets partially blanked out by the fuselage, and the down left aileron increases the angle of attack on that portion of the wing which, of course, results in an earlier stall. In these cases, the ball is deflected in the direction of the roll, but the aircraft will still roll away from the ball as it stalls.

Try to remember that you have control of the aircraft during the stalls. If it does something unexpectedly, relax and analyze the situation. Remember, *you're* the one who caused it to happen!

LANDINGS

The normal landing for a private student during solo is broken down into three phases:

- ☐ Prior to the Stage One Check—no flaps.
- ☐ After the Stage One Check until certified by an instructor—maximum 20 degree flaps.
- ☐ After certification by the instructor—full flaps.

A normal approach to landing is started on the downwind leg by applying carburetor heat about halfway down the runway to eliminate any ice formation. Opposite the point of intended landing, the throttle is smoothly reduced to approach power (1500-1700 rpm). Now we must hold the nose up so the aircraft will decelerate to the approach speed (65 kts) without losing any altitude. After the airspeed has decayed to the approach speed, the nose can be slightly lowered to prevent further loss of airspeed and the aircraft trimmed for hands-off.

When the end of the runway appears to be about 45 degrees behind you, it is time for the turn to base. The wing must be lifted prior to the turn so we can see if any traffic is turning inside you or overtaking you from that side. A good rule is that 30 degrees of bank is the maximum to use while in the pattern and 20 to 25 degrees will produce a comfortable rate of turn. If the turn is too shallow, we are blind for a long period of time and the radius of turn will be too great, which will cause us to overshoot the centerline of the runway. As we roll out of the bank for the turn to base, the lower wing should be raised above level to immediately check for traffic turning inside you. Once this is accomplished, we are at the "key position," where we can determine if any large changes in power must be used to adjust our angle of descent. After clearing the area to the right for any traffic on final for a straight in approach, we can begin our turn to final leg.

Don't be afraid to lean forward in the seat and watch the runway through the windshield so you can line up on the extended centerline of the runway. Here, slight variations of bank will increase or decrease the radius of turn so we can follow our desired ground track. Still keeping the airspeed constant, we can watch the runway; if it appears to be coming under us, we are overshooting and must reduce power to increase our angle of descent. If the touchdown area does not appear to move but only gets larger, we are on the correct glide angle, but if it appears to move up, we are too low and must add power.

Remember that if the airspeed remains constant, power controls the altitude. As we cross the end of the runway, start a slow, smooth flare to stop the rate of descent. As the aircraft slowly decelerates, continue to increase the pitch attitude to prevent it from settling onto the runway at an excessive groundspeed. During this transition period our eyes should move out in front of the aircraft to the same position as they would be if you were driving your car down a highway at 55-60 mph. If you look too close to the aircraft, there will be a tendency to flare too high, whereas if you look too far down the runway, you will not be able to detect any change in altitude or height above the runway, and will tend to flare too late.

We can keep raising the nose (as long as the aircraft does not climb) until we reach the landing attitude. If we maintain this attitude by slowly continuing the back pressure, the aircraft will slowly settle to the runway as it continues to decelerate. After the main wheels are on the ground, maintain the same pitch attitude with the elevator to prevent the nosewheel from coming down until the aircraft gets so slow that you can no longer hold the nose off and it will also slowly settle to the runway.

It is important to realize that we really don't land the aircraft, but only reach a given attitude and power combination at a specific height above the runway and the aircraft lands itself. While we are approaching this attitude and height, we must also keep the aircraft flying down the centerline of the runway and keep the longitudinal axis or heading parallel with the runway to prevent any side loads on the landing gear.

Landing with Flaps

With 20 degrees of flaps, the only things that change are the actual extending of the flaps and the approach speed and attitude. When we reduce our power to approach power (1500-1700 rpm) and hold the nose up to slow down to our approach speed, we can add our 20 degrees of flaps 10 degrees at a time. As the flaps extend, the aircraft tends to pitch up, so it will be our job to maintain the "slowing down" pitch attitude. The pitching tendency of the aircraft while adding the first 20 degrees of flaps is the primary reason why we want this accomplished before the descent is initiated.

Now the airspeed is stabilized at 60 kts,

the power is constant, the flaps are on, and the aircraft is trimmed before the return to base leg is initiated. It sounds like there is much more to do now, but there really isn't because very little if any trim change must be made. As the power is reduced we must again hold the nose up to start decelerating; as this is happening we will put 10 degrees of flaps on. While these flaps extend, we will have to reduce some back pressure and let the aircraft stabilize momentarily. Then the flaps may be extended to 20 degrees and again release some back pressure to prevent the pitching up. With the 20 degrees of flaps on, the pitch attitude to maintain our 60 kts will be considerably lower than without flaps and 65 kts; however, there is very little trim change between cruise flight (or downwind) and approach power, speed, and 20 degrees flaps. Again we will allow the aircraft to stabilize its airspeed and begin its descent before turning onto base.

With flaps extended, the glide angle will be a little steeper than without flaps, so as you reach the key position you will appear to be high, but wait and see if the touchdown area is moving before any power adjustments are made. Again, with the constant airspeed, the altitude is controlled with throttle and the pattern is continued to the final leg. The throttle should be completely closed by the time you cross the threshold of the runway and the flare is again started slowly and smoothly. The landing attitude of the aircraft does not change from that of no flaps, but the ground speed will be considerably less due to the increased lift from the flaps. This is a safer landing because the touchdown speed is lower. If a touch-and-go landing is being used, the flaps must be checked in the full up position by looking at the flaps rather than the indicator before the throttle is advanced for another takeoff.

A full flap landing carries the 20 degree flap techniques one step further. Twenty degrees are still put on during the downwind leg, but the turn to base may be started a little sooner due to the increased rate and angle of descent as more flaps are added. After the key position, if we determine that we are indeed too high, more flap can be added. Now the nose must be lowered slightly to maintain our speed as we turn onto final. If we see that we are too high on final, the last flaps are added; now if we are still too high, the power may be reduced.

Notice that 1500-1700 rpm was maintained until all the flaps were extended, and flaps were not extended beyond 20 degrees until we noticed we are high. This is to prevent you from getting low with full flaps. In most aircraft, a low approach with full flaps is a dangerous situation in that any shortage of power with such extreme drag from the flaps may put you short of the runway. The only difference in the actual landing will be the ground speed—the attitude will be the same. Again, if you are making a touch and go, the flaps will be visually checked in the full up position before the throttle is advanced for the takeoff.

Forward Slips to a Landing

These are not normally used in the Cessna 152, but after you get your license, you are not limited to flying the 152. Therefore, you will get an introduction to slips during the private phase of training.

The main reason for a slip to a landing is to lose altitude more rapidly without gaining airspeed. We do this in a 152 by adding flaps, but some other aircraft are not equipped with flaps, so a slip would be needed to lose altitude. One thing to remember in a slip is that the airspeed indicator is *not* reliable. The airspeed indicator does nothing more than measure the difference in pressure between the pitot tube and the static source. In a left slip, the static port is exposed partly to ram air so the pressure differential in the instrument is less and the airspeed will read low, whereas in a right slip, the static port is exposed to a lower pressure and the airspeed will read high.

The slip is entered after turning onto a high final by lowering one wing with aileron and then applying opposite rudder to stop the aircraft from turning. What we want is the turning force from the wings to be counteracted by the rudder so the aircraft approaches the runway slightly sideways, but the ground track is still down the centerline of the runway. Since the aircraft is going through the air somewhat sideways, there is a considerable increase in drag and this will in turn increase the rate and the angle of our descent. Remember to fly the pitch attitude outside and not the airspeed indicator. Prior to touchdown, you can relax the aileron and rudder pressures and the aircraft will straighten out for a normal or no-flap landing.

Crosswind Landings

The simplest way for the student to make a crosswind landing is by the *sideslip* method. During a sideslip, the longitudinal axis of the aircraft remains parallel with the runway centerline. After rolling out on final approach, you can determine if there is a crosswind by noting the crab needed to hold your ground track down the centerline of the runway. If the aircraft has to be crabbed to the left, the crosswind is from the left; if the wind is from the right, the aircraft must be crabbed to the right. If we were to align the aircraft with the runway, the wind would drift it off to the downwind side. Our problem with a crosswind is twofold: Align the longitudinal axis with the runway, and prevent the sideward drifting. For the remainder of this narrative, we will consider a crosswind from the left.

While on short final, we will lower the left wing and then apply right rudder. With the wing down, it will produce some horizontal lift to offset the wind drift, but since the aircraft will turn when it is banked, we must stop the turn by using the opposite (right) rudder. Think of the rudder as doing whatever is necessary to align the aircraft with the runway. Aileron controls drift; rudder keeps the aircraft straight. Since the aircraft is banked, the left main wheel will touch first. Don't' let this alarm you because damage cannot occur if there is no sideward force. Now we will continue to add aileron as the aircraft slows down until we have full aileron into the wind. More rudder will also be needed to keep the aircraft straight on the rollout. Remember that if some power is left on, the propeller wash will keep the rudder and elevator more effective at slower airspeeds.

After full aileron is reached and the aircraft continues to decelerate, the right main wheel will come down. Note that it is important to gently lower the nosewheel before the other main wheel comes down. This will prevent a high angle of attack on the aircraft before running out of aileron control.

The most common errors are landing the aircraft at higher-than normal speeds, and putting the wing down to a certain bank and not changing it even if the crosswind varies. The pitch attitude in smooth crosswinds will be very close to a normal landing without a wind. As a student, it is also important to recognize your capabilities in a crosswind. Undoubtedly, your ability to handle winds from 90 degrees to the runway will be less than 25 kts. If there is a 35 kt wind directly down the runway, this shouldn't give you any problem during the landing, but it is *dangerous* to turn the aircraft *out of the wind*. Anytime you run into a situation where the wind is more than 25 kts, you should call up the flight school or facility where you landed and ask for wing walkers! Once the wind gets under the wing and starts lifting it, things will happen so fast the next thing you know is that you're on your back, hanging from the seat belt! ALWAYS RESPECT THE WIND AND KNOW YOUR OWN LIMITATIONS.

During gusty crosswinds, the approach speed on final should be slightly higher than normal to maintain better control of the aircraft. Now we must also dissipate this excess speed before touchdown. Touchdown will be the same as above, but the nosewheel should be lowered a bit sooner to help with directional control.

It is impossible to fully explain when and where all these things occur, but after doing a few with your instructor, you will soon realize your limitations and capabilities. Usually, once the student is on the ground, the thought occurs that the problems are all over. Nothing is further from the truth! The most critical time in a crosswind landing is during the rollout phase after touchdown and before reaching taxi speed. And you are never really through flying until the aircraft is tied down!

Short Field Takeoff and Landing

The short field takeoff assumes that there is at least a 50 foot obstacle near the end of the runway that must be cleared immediately after takeoff. This takeoff is special because we want the most altitude in the horizontal distance we have available, and for that we must use the best angle of climb speed or V_x. The takeoff roll is started from the extreme end of the runway to use all the available distance.

After the throttle is smoothly and fully opened, and the power check is made, the aircraft is allowed to accelerate to V_x minus 5 and then smoothly rotated to the attitude that will maintain V_x. After a few times this attitude will be familiar to you, but it will seem a bit steep at first. The attitude and airspeed are held until the obstacle is cleared and then the nose is lowered carefully to allow the aircraft to accelerate to its best rate of climb speed without losing altitude. After reaching a safe altitude, accelerate to the cruise climb speed.

The short field landing technique assumes that you are attempting to land the aircraft on a relatively short runway and there is a 50′ obstacle at the approach end of the runway. We have two problems: We must get the aircraft onto the runway as close to the approach end as possible, and we must get the aircraft stopped as soon as possible. For training purposes, we will not do a maximum performance stop due to the higher maintenance costs of such operations, but the techniques can still be applied and developed fully.

In the aircraft owner's handbook a speed, configuration, and power setting is recommended and this is what we'll follow. A normal approach is started and continued until we turn onto final, where we will get out full flaps on and raise the nose slightly to reduce the airspeed to the desired short field approach speed. Like any approach, we're looking for an angle of descent that will take us to a desired touchdown point, but this time the angle is steeper due to a slower speed.

If we see that we are overshooting, reduce the power as necessary to regain the angle. If we are low, add power. By the time we are over the simulated obstacle the power should be set and the descent is continued to a normal flare and touchdown. Assuming the airspeed is correct, the power will be reduced to idle as we start the flare for landing. If the airspeed is too slow, we may have to carry power until after the flare is started. Note that this is a slightly different technique than what is spelled out in the C-152 manual, but it's the FAA recommended short field landing. They don't want you to pull the power and lower the nose over the obstacle because of the high descent rate that will result.

After the mains are on the runway, the nosewheel is lowered to the ground so that brakes can be applied and the flaps are retracted to put more weight on the wheels so the braking is more effective. Again, this will not be carried through to a maximum performance stop. Care should be taken not to slam the nosewheel onto the ground; brakes are not applied until the nosewheel is on the ground. The hardest part of this approach is judging the angle, but once you see how to judge if you're high or low, it should be easy.

A short field landing is a very precise maneuver and the best thing you can have going for you is airspeed control. If the airspeed is too fast, you will float excessively after the flare, whereas if the airspeed is too slow, you will not be able to stop the descent with flare alone and this will probably result in a hard landing.

Soft Field Technique for Takeoff

The soft field takeoff technique should be used on any runway or field that does not have a hard, smooth surface. There are some paved runways that are actually rough enough to require soft field techniques to minimize any chance of damage to the aircraft.

The most fragile wheel of a tricycle-geared airplane is the nosewheel, so that's the one you want to protect the most. In mud, snow, tall grass, or a rough field, the drag on the wheels tends to nose the aircraft over and can also lengthen the takeoff roll considerably, so we want to become airborne as soon as possible to get rid of all the drag. In the C-152 handbook, the factory recommends that 10 degrees flaps be used to produce more lift at slower speeds.

As you roll onto the runway, the flaps are already extended and it is important to keep the aircraft moving and hold the elevator full back to lighten the load on the nosewheel. If the aircraft is stopped on a muddy or snow-covered runway, it will be extremely difficult to get it moving again. It takes only about two inches of obstruction at the wheel to neutralize the available thrust from the propeller. Even if you could get the aircraft moving again, damage to the propeller could easily result from picking up ice, rocks, or trash from the ground.

So keep it rolling onto the runway with 10 degrees flap and full back elevator, and apply full power smoothly as you line up on the centerline. Remember that since the elevator and rudder gain much of their effectiveness from the propwash, as you add power, the nose will want to pitch up. As it does you must relieve a little back pressure to avoid banging the tail on the runway. This pitch-up happens after the aircraft gains enough speed to make the elevator effective enough to raise the nosewheel.

Steering is now accomplished entirely with the rudder. At this slow airspeed and high angle of attack, we must use considerable right rudder to compensate for torque and P-factor. As the aircraft continues to accelerate, the elevator gains effectiveness and back pressure must be slowly released to keep the same attitude.

When the airplane is ready, it will fly off the runway in a higher than normal nose-high attitude. Now substantial back pressure must be released to avoid an excessively nose-high attitude as the aircraft leaves ground effect. We want to establish the best rate of climb airspeed (V_y) before actually climbing so the nose must be slowly lowered until the the proper speed is reached. Flaps should be retracted only after V_x has been reached.

The Soft Field Landing Technique

The soft field landing technique should be used whenever the runway is not a hard, firm surface. Good examples are snow or slush on a paved runway, or after heavy or steady rains on grass or gravel runways. It is used primarily to decrease the possibility of damaging the aircraft as the wheels sink into the soft surface.

The idea is to land the aircraft as slowly as possible and to touch down as smoothly and softly as possible. Since the airplane can fly slower with flaps on than without, full flaps are used. Also, the airplane will fly slower with power than without, so approximately 1800-2000 rpm is used; any more than that and the aircraft will continue to fly in the ground cushion, and if any less is used the aircraft will not be flying as slowly as possible when it lands.

The soft field landing is started with a normal approach planning full flaps on final. After the flaps are fully extended, the power is reduced to idle as we see that the runway is made, just as in a normal approach. The same flare is also used, but from here on, things get different.

After the aircraft is flared to stop the rate of descent, we continue to slowly increase the pitch attitude to get more angle of attack (lift) because the strength of the relative wind is decreasing. This process is slowly continued until the desired pitch attitude for landing is reached.

Do not increase the pitch attitude or the tail will strike the ground first. Power is slowly and smoothly added back in to maximum of about 2000 rpm. As the power is applied, it will become necessary to release some back pressure to maintain the same attitude and height above the runway, and we must be careful not to let the aircraft start flying again ("balloon" back up). Now it is just a matter of controlling the power and pitch attitude to keep the aircraft from climbing or settling onto the runway prematurely. Remember, the aircraft will decelerate much slower with the power on, so this may seem like quite a drawn-out affair compared with a normal landing.

As the aircraft slowly settles onto the runway and continues decelerating, the nosewheel is held off as long as possible and then the throttle is slowly closed. Remember that the primary way to land the aircraft on a soft field is as slowly as possible and to touch down as softly as possible. Another advantage of using power during the actual landing process is that the propwash makes the elevator more effective, thereby allowing you to hold the nosewheel off even longer and with more control. Also, some of the thrust from the propeller is now acting vertically to lessen the weight on the wings and help transfer the load to the wheels as gently as possible. Power will also lower your stall speed to below the published V_{so}, since V_{so} was determined with power off.

Appendix

Index of Federal Aviation Regulations

FAR PART 61 CERTIFICATION OF PILOTS AND FLIGHT INSTRUCTORS

Subpart A General

61.1	Applicability
61.3	Requirements for certificates and ratings
61.5	Certificates and ratings issues
61.15	Offenses involving narcotic drugs
61.19	Duration of pilot and C.F.I. certificates
61.23	Duration of medical certificates
61.25	Change of name
61.27	Voluntary surrender or exchange of certificate
61.29	Replacement of lost or destroyed certificate
61.31	General limitations
61.33	Tests; general procedure
61.35	Written test; prerequisites and passing grades
61.37	Written tests; cheating or unauthorized conduct
61.39	Prerequisites for flight tests
61.49	Retesting after failure
61.51	Pilot logbooks
61.53	Operations during medical deficiency
61.57	Recent flight experience; pilot in command
61.60	Change of address

Subpart C Student Pilots

61.85	Application
61.87	Requirements for solo flight
61.89	General limitations
61.91	Aircraft limitations; pilot in command
61.93	Cross-Country flight requirements
61.101	Applicability
61.103	Eligibility requirements; general
61.105	Aeronautical knowledge
61.107	Flight proficiency
61.109	Airplane rating; aeronautical experience
61.118	Private pilot privileges and limitations

FAR PART 71 CONTROLLED AIRSPACE

71.7	Control areas
71.9	Continental control areas
71.11	Control zones
71.12	Terminal control areas
71.13	Transition areas

FAR PART 91 GENERAL OPERATING AND FLIGHT RULES

91.1	Applicability
91.3	Responsibility & authority of the p. i. c.
91.5	Preflight action
91.7	Flight crewmembers at stations
91.8	Prohibition against interference with crewmembers
91.9	Careless or reckless operation
91.10	Careless or reckless operation other than for the purpose of air navigation
91.11	Liquor and drugs
91.12	Carriage of narcotic drugs and substances
91.13	Dropping of objects
91.14	Use of safety belts
91.22	Fuel requirements for flight under VFR
91.24	ATC transponder and altitude reporting equipment
91.27	Civil aircraft; certifications required
91.29	Civil aircraft airworthiness
91.31	Civil aircraft operating limitations and markings
91.32	Supplement oxygen
91.33	Instrument and equipment requirements
91.39	Restricted category civil aircraft; operating limitations

91.52 Emergency locator transmitters

Subpart B Flight Rules

91.61 Applicability
91.65 Operating near other aircraft
91.67 Right-of-way rules; except water operations
91.70 Aircraft speed
91.71 Acrobatic flight
91.73 Aircraft lights
91.75 Compliance with ATC clearances and instructions
91.77 ATC light signals
91.79 Minimum safe altitudes; general
91.81 Altimeter settings
91.83 Flight plan; information required
91.84 Flights between Mexico or Canada and the U.S.
91.85 Operating on or in the vicinity of an airport
91.87 Operations with operating control towers
91.89 Operation at airports without control towers
91.90 Terminal control areas
91.91 Temporary flight restrictions
91.93 Flight test areas
91.95 Restricted and prohibited areas
91.97 Positive control areas and route segments
91.101 Operations to or over Cuba
91.102 Flight limitation near space flight recoveries
91.103 Operation of civil aircraft of Cuban registry
91.104 Flight restrictions near the Presidential party
91.105 Basic VFR weather minimums
91.107 Special VFR weather minimums
91.109 VFR cruising altitude or flight level

Subpart C Maintenance, Preventive Maintenance, and Alterations

91.161 Applicability
91.163 General
91.165 Maintenance required
91.167 Carrying persons after repairs or alterations
91.169 Inspections
91.173 Maintenance records

NATIONAL TRANSPORTATION SAFETY BOARD

830 Notification & reporting of accidents & incidents

Index

Edited by Steven H. Mesner